Studies in Fuzziness and Soft Computing

Volume 318

Series editor

Janusz Kacprzyk, Polish Academy of Sciences, Warsaw, Poland
e-mail: kacprzyk@ibspan.waw.pl

For further volumes:
http://www.springer.com/series/2941

About this Series

The series "Studies in Fuzziness and Soft Computing" contains publications on various topics in the area of soft computing, which include fuzzy sets, rough sets, neural networks, evolutionary computation, probabilistic and evidential reasoning, multi-valued logic, and related fields. The publications within "Studies in Fuzziness and Soft Computing" are primarily monographs and edited volumes. They cover significant recent developments in the field, both of a foundational and applicable character. An important feature of the series is its short publication time and world-wide distribution. This permits a rapid and broad dissemination of research results.

Peter C. Casey · Michael B. Gibilisco
Carly A. Goodman · Kelly Nelson Pook
John N. Mordeson · Mark J. Wierman
Terry D. Clark

Fuzzy Social Choice Models

Explaining the Government Formation Process

Springer

Peter C. Casey
Department of Political Science
Washington University in St. Louis
St. Louis, MO
USA

Michael B. Gibilisco
Department of Political Science
University of Rochester
Rochester, NY
USA

Carly A. Goodman
West Corporation
Creighton University
Omaha, NE
USA

Kelly Nelson Pook
Terry D. Clark
Department of Political Science
Creighton University
Omaha, NE
USA

John N. Mordeson
Department of Mathematics
Creighton University
Omaha, NE
USA

Mark J. Wierman
Computer Science and Informatics
Creighton University
Omaha, NE
USA

ISSN 1434-9922   ISSN 1860-0808   (electronic)
ISBN 978-3-319-35673-0   ISBN 978-3-319-08248-6   (eBook)
DOI 10.1007/978-3-319-08248-6
Springer Cham Heidelberg New York Dordrecht London

Printed on acid-free paper

Springer is part of Springer Science+Business Media (www.springer.com)

We dedicate this book to Rose Hill, who is always there when you need her

Preface

John N. Mordeson, Mark J. Wierman, and Terry D. Clark began working 8 years ago on the application of fuzzy Mathematics to public choice models. Their initial intent was to explain the formation of governments in parliamentary systems. However, they soon discovered that they would have to devote significant attention to matters of theory. As a consequence, they ended up producing what can only be described as a very large volume of work on public choice theory. While they also produced some empirical work, the best of which has appeared on the pages of the journal *Public Choice*, the theoretical work has clearly overshadowed the empirical work in sheer volume.

This book partially redresses that imbalance. It is a compilation of most, if not all, of the effort that went into the empirical question that motivated the initial project. In it we present the results of several attempts to predict the outcome of the government formation process in parliamentary systems using fuzzy public choice models. However, even in this volume we present a substantial amount of the theoretical work related to the fuzzy models. While much of it has appeared previously in print, most readers would find understanding the approach that we take with each model incomprehensible had we not restated many of our main theoretical findings here.

As those who are familiar with our work are aware, we have engaged many bright young students over the last 8 years. Several of them have gone on to pursue Ph.D. in Political Science or Mathematics. This is as true for our empirical work as it is for our theoretical work. Peter Casey and Michael Gibilisco took the lead in the work presented here. They were assisted by a very large number of their student colleagues, one of whom Carly Goodman was particularly instrumental in helping to formulate the approaches and undertake the tests that led to the results that we report in the book in front of you. Carly also took the lead in editing the resulting papers, which included checking the models and re-verifying the results. Kelly Pook finished the task, which Carly passed on to her before departing for a lucrative position as a business analyst.

Peter C. Casey, is presently pursuing Ph.D. in Political Science at the University of Washington in St. Louis. Peter dedicates this book to his mother, Virginia Casey, for her guidance and support. Michael Gibilisco is pursuing Ph.D. in Political Science at the University of Rochester. Michael dedicates this book to his parents whose moral, and, at times, financial support, made the work possible.

They have always encouraged him and his research throughout school and this project, and his passion for learning began with them. Carly Goodman is grateful to her co-authors and mentors in the Fuzzy Research Colloquium, without whom this book would not have been possible. She dedicates her contribution to her parents and to Eric Norrgard for their constant encouragement and support. Kelly Pook dedicates her work to Ryan, for the first year of many. She is at present the head research assistant for the Social Network Analysis Working Group. As such, she supervises the work of 20 students. John Mordeson dedicates this book to his loving wife Pat. Mark J. Wierman dedicates this book to Mary K. Dobransky. Terry D. Clark dedicates his work in this book to his wife of 37 years, Marnie, whom he adores, and to his granddaughter, Zoey, whose coming into this world has brought such joy.

Omaha, USA, March 2014   Terry D. Clark
                                                                                     John N. Mordeson
                                                                                     Mark J. Wierman

Acknowledgments

This research grew out of the Fuzzy Spatial Modeling Colloquium. The colloquium is indebted to Prof. Bridget Keegan, Interim Dean of the College of Arts and Sciences at Creighton University whose support has been invaluable in sustaining our efforts.

We are also indebted to Dr. George Haddix and his late wife Sally Haddix for their generous endowments to the Department of Mathematics at Creighton University.

Finally, we thank the journal, New Mathematics and Natural Computation, for its support of research using fuzzy Mathematics in Political Science and for allowing us to reuse some of our work for this book.

Contents

Chapter 1
A Fuzzy Public Choice Model

Abstract Public choice models can be a powerful tool for the explanation and prediction of political phenomena in the discipline of political science. However, the predictions of such models depend to a considerable extent on whether the preferences of political actors can be properly estimated. Fuzzy mathematics offers one possibility for dealing with this challenge. A fuzzy approach permits us to consider the preferences of individuals as ambiguous, marked by a considerable degree of indifference related to actors' uncertainty about the exactitude of their ideal points. Moreover, fuzzy sets also permit us to reconsider approaches to making predictions on the basis of those fuzzy preferences. One particularly interesting possibility is offered by a fuzzy maximal set.

1.1 Improving the Prediction of Public Choice Models: A Fuzzy Approach

The extraction of preference measures from empirical data has been an on-going challenge in the verification of public choice models. Beyond the inherent theoretical problems of rationalizability,[1] which demonstrates that players' preferences may not be observable, lies the problem of inferring players' preferences from observable actions, should such observations even be possible. Among these problems lies the selection of appropriate raw data, selecting the method to extract measures from the data, avoiding systematic error (bias), and measuring nonsystematic (unbiased) error.

Laver (2001) provides an excellent overview of these and other issues related to the estimation of preferences. He devotes particular attention to the use of Comparative Manifesto Project (CMP) data. CMP data provide one of the most widely available and exhaustive sources of text-derived preferences. They contain manifestos

---

[1] The problem of rationalizability is the problem that the preferences of political actors cannot be inferred from the behaviors those preferences motivate. A discussion of the problem of rationalizability can be found in Austen-Smith and Banks (1999).

P. C. Casey et al., *Fuzzy Social Choice Models*, Studies in Fuzziness and Soft Computing 318, DOI: 10.1007/978-3-319-08248-6_1, © Springer International Publishing Switzerland 2014

for some 780 political parties released prior to 3,018 elections in 54 European countries spanning the period 1945–2005 (Budge et al. 2001; Klingemann et al. 2006). The manifestors are parsed into "quasi-sentences"—or phrases that reflect complete thoughts—by CMP coders and placed numerically into one of 56 policy dimensions, or designated as "unknown" in the case of insufficient data (Budge et al. 2001; Klingemann et al. 2006). Insufficient or missing data are addressed by collecting and analyzing supplemental party materials. The policy dimensions can be aggregated to create various scalars, or *riles*, along which parties can be placed in relation to one another. One of the most popular of these is the left-right political spectrum. Laver and Budge (1992, pp. 22–30) place parties in relation to one another on this spectrum by taking the difference of the sum of the number of quasi-sentences considered as being right-wing and left-wing respectively.

Scholars using CMP data have expressed some concern about the reliability of determining preferences from text data. Benoit et al. contend that the stochastic process of writing the text, or text generation, introduces non-systematic (unbiased) error into preference measures. Language is inherently vague, and it is possible that, in the drafting and re-drafting of a political text, an author may not represent his preferences "precisely". Thus, any attempt to derive the preferences of a political actor from a political text will be affected by some random error.

Benoit et al. (2009) use a bootstrapping procedure in an attempt to replicate this stochastic process and calculate error scores for the CMP data. Their approach to estimating preferences from CMP data involves taking the total number of quasi-sentences in a manifesto and their assigned categories and randomly reassigning the quasi-sentences to each category. The process is repeated 1,000 times to produce a standard error of policy uncertainty, which is taken to represent the error, or *noise*, in the CMP data.

Benoit et al. defend the use of the bootstrapping procedure by arguing the authors of political texts have a precise policy position that they can communicate accurately through the proper allocation of quasi-sentences. However, through the stochastic process of text generation, these authors may accidentally misallocate quasi-sentences to the different policy categories they mention, thus overemphasizing some policies and underemphasizing others, communicating a policy position other than their intended message. Therefore, the policy position communicated by the text may be another policy position on the relevant dimension with some random error differentiating it from the author's actual position.

The approach taken by Benoit et al. attempts to determine the precise policy preference of political parties using what are arguably inherently vague documents. In our view, there is substantial reason to believe that parties do not wish to be overly precise in communicating their policy preferences to either potential candidates, voters, or the opposition. Moreover, the ambiguity of the documents is increased by the very nature of language as an instrument for communication.

The effort we report in this book, applies fuzzy mathematics to the CMP data with the intent of improving the capacity of public choice models to predict the outcome of the government formation process in parliamentary systems. We take the standard rationalist approach to the problem by assuming that parties whose policy preferences are most closely aligned are more likely to agree to a government coalition. However, we eschew the conventional approach of attempting to locate those policy preferences in crisp Euclidean space. Rather, we take advantage of fuzzy mathematics ability to deal with uncertainty over the exactitude that players themselves have about their policy preferences.

Fuzzy mathematics is intended to deal with vagueness and ambiguity. Thus, in our view, the so-called "error" revealed by the bootstrap procedure represents uncertainty in the party's policy preferences. Instead of having a single, crisp ideal point, parties preferences encompass regions within Euclidean space. These regions represent areas where parties can find common ground with one another, often with little to no loss of utility. As we develop our argument, we will refer to these as fuzzy preference profiles. The goal of our book is to use fuzzy preference profiles in public choice models to demonstrate the resulting enhanced capacity to predict the outcomes of the government formation process in European parliamentary systems.

In the remainder of this chapter, we discuss the basic concepts of fuzzy mathematics. We conclude the chapter by introducing the fuzzy maximal set. These sets will be quite important in some of the efforts we present in later chapters that attempt to predict the outcome of the government formation process.

In Chap. 2, we begin by presenting a method for extracting fuzzy preference measures from CMP data using the bootstrap procedure designed by Benoit et al. (2009) to calculate error in the CMP data. We use the bootstrapping method to project a distribution of the possible outcomes of text generation. We then use this distribution of policy positions to determine the bounds and alpha-levels of the fuzzy numbers representing each party's preference profile. Essentially, we treat uncertainty as 'noise' that eliminates the CMP's ability to identify precise policy positions, and use that uncertainty concerning a party's preferences to map how the party perceives shifts in utility over a subset of policy space.

We conclude Chap. 2 by discussing how fuzzy preferences can be used in public choice models. The resulting fuzzy public choice model can explain seemingly detrimental shifts in policy on one dimension in favor of gains on another dimension. In conventional public choice models, a political actor will only give way to a loss in utility on one dimension if it entails an overall gain in utility in multidimensional space. A fuzzy public choice model, on the contrary, allows an actor to make compromises on one dimension for gains in another without necessarily experiencing a loss in utility on the former. In this way, scholars can predict whether an actor may be more likely to make a compromise for a gain, when the actor perceives no loss in utility on the compromised dimension, while the actor may be more resistant to such a trade-off if some real loss in utility occurs as a result. Thus we see that the fuzzy public choice model satisfies some of the intuitive and practical problems faced by the conventional model.

In Chap. 3, we consider mathematical issues related to fuzzy single-dimensional public choice models. We give particular attention to Black's Median Voter Theorem, which argues that political actors will collectively prefer the median alternative if the set of alternatives can be arranged on a single dimension in such a manner that preferences of all actors over other alternatives descend monotonically from their ideal point.

In Chap. 4, we make use of the concepts developed in the first three chapters to present a fuzzy single-dimensional public model to predict the outcome of the government formation process. The first model makes use of the fuzzy maximal set, a second model considers the utility of fuzzy Pareto sets for predicting outcomes. Using our method for extracting fuzzy preferences from CMP data, we compare the predictions made by both models with data about the actual governing coalitions formed following an election. In Chap. 5, we turn to a consideration of the mathematical issues related to fuzzy two-dimensional public models. Based on the discussion, we present a fuzzy two-dimensional model for predicting the outcome of the government formation process in Chap. 6. Chapter 7 concludes with a brief consideration of weighted models and new directions for using fuzzy public choice models.

We now turn to a preliminary discussion of the basics of fuzzy mathematics. We conclude by introducing the fuzzy maximal set.

1.2 Sets

We begin by introducing basic fuzzy set notation and concepts. These concepts have important applications in public choice models.

Let S be a set and let A and B be subsets of S. We use the notation $A \cup B$ and $A \cap B$ to denote the union and intersection of A and B, respectively. We also let $A \setminus B$ denote the relative complement of A in B. The relative complement of A in S, $S \setminus A$, is sometimes denoted by A^c when S is understood.

It is easily verified that $(A \cup B)^c = A^c \cap B^c$ and $(A \cap B)^c = A^c \cup B^c$. These equations are known as DeMorgan's Laws.

Let x be an element of S. If x is an element of A, we write $x \in A$, otherwise we write $x \notin A$.

We use the notation $A \subseteq B$ or $B \supseteq A$ to denote that A is a subset of B. If $A \subseteq B$ and there exists $x \in B$ such that $x \notin A$, then we write $A \subset B$ or $B \supset A$ and we say that A is a proper subset of B.

The cardinality of A is denoted by $|A|$ and is the number of elements that are contained in X.

The power set of A, written $P(A)$, is defined to be the set of all subsets of A, i.e., $P(A) = \{B \mid B \subseteq A\}$. In particular, we let $\mathscr{P}(X)$ denote the power set of X, i.e., the set of all subsets of the universal set X.

We let $\mathbb{N}$ denote the set of positive integers, $\mathbb{Z}$ the set of integers, $\mathbb{Q}$ the set of rational numbers, and $\mathbb{R}$ the set of real numbers.

Let X and Y be sets. If $x \in X$ and $y \in Y$, then (x, y) denotes the ordered pair of x with y. The Cartesian cross product of X with Y is defined to be the set $\{(x, y) \mid x \in X, y \in Y\}$ and is denoted by $X \times Y$. At times we write X^2 for $X \times X$. In fact, for $n \in \mathbb{N}$, we let X^n denote the set of all ordered n − tuples of elements from

X. A relation R of X into Y is a subset of $X \times Y$. Let R be such a relation. Then the domain of R, written $\text{Dom}(R)$, is $\{x \in X \mid \exists y \in Y \text{ such that } (x, y) \in R\}$ and the image of R, written $\text{Im}(R)$, is $\{y \in Y \mid \exists x \in X \text{ such that } (x, y) \in R\}$. If $(x, y) \in R$, we sometimes write $x R y$ or $R(x) = y$. If R is a relation from X into X, we say that R is a relation on X.

Definition 1.1 A relation R on X is called

1. **reflexive** if $\forall x \in X, (x, x) \in R$;
2. **symmetric** if $\forall x, y \in X, (x, y) \in R$ implies $(y, x) \in R$;
3. **transitive** if $\forall x, y, z \in X, (x, y), (y, z) \in R$ implies $(x, z) \in R$.

A relation R on X is called antisymmetric if $\forall x, y \in X, (x, y) \in R$ and $(y, x) \in R$ implies $x = y$. If R is a reflexive, antisymmetric, and transitive relation on X, then R is called a partial order on X and X is said to be partially ordered by R. A binary relation R on X is called **complete** if for all $x, y \in X$, either $(x, y) \in R$ or $(y, x) \in R$.

Let R be a relation of X into Y, and T a relation of Y into a set Z. Then the composition of R, with T written $T \circ R$, is defined to be the relation $\{(x, z) \in X \times Z \mid \exists y \in Y \text{ such that } (x, y) \in R \text{ and } (y, z) \in T\}$. If f is a relation of X into Y such that $\text{Dom}(f) = X$ and $\forall x, x' \in X, x = x'$ implies $f(x) = f(x')$, then f is called a function of X into Y and we write $f : X \rightarrow Y$. Let f be a function of X into Y. Then f is sometimes called a mapping of X into Y. If $\forall y \in Y, \exists x \in X$ such that $f(x) = y$, then f is said to be onto Y or to map X onto Y. If $\forall x, x' \in X, f(x) = f(x')$ implies $x = x'$, then f is said to be one-to-one and f is called an injection. If f is a one-to-one function of X onto Y, then f is called a bijection. If g is a function of Y into a set Z, then the composition of f with g, $g \circ f$, is a function of X into Z which is one-to-one if f and g are one-to-one and which is onto if f onto Y and g is onto Z. If $\text{Im}(f)$ is finite, then f is called finite-valued. We say that an infinite set X is countable if there exists a one-to-one function of N onto X, otherwise X is called uncountable.

We use the notation $a \vee b$ to denote maximum of a and b and $a \wedge b$ to denote the minimum of a and b.

1.3 Fuzzy Sets

A fuzzy subset of X is a function of X into the closed interval $[0, 1]$. We use the notation $\mu : X \rightarrow [0, 1]$ to denote a fuzzy subset μ of X. We let $\mathscr{F}\mathscr{P}(X)$ denote the fuzzy power set of X, i.e., the set of all fuzzy subsets of X. If $\mu \in \mathscr{F}\mathscr{P}(X)$, we define the support of μ, written $\text{Supp}(\mu)$, to be $\{x \in X \mid \mu(x) > 0\}$. Let $\mathscr{P}^*(X) = \mathscr{P}(X) \backslash \{\emptyset\}$ and $\mathscr{F}\mathscr{P}^*(X) = \{\mu \in \mathscr{F}\mathscr{P}(X) \mid \text{Supp}(\mu) \neq \emptyset\}$. If S is a subset of X, we let 1_S denote the characteristic function of S in X. That is, 1_S is the function of X into $\{0, 1\}$ such that $1_S(x) = 1$ if $x \in S$ and $1_S(x) = 0$ otherwise. We let θ denote the fuzzy subset of X such that $\theta(x) = 0$ for all $x \in X$. Let $\mu, \nu \in \mathscr{F}\mathscr{P}(X)$. Then we write $\mu \subseteq \nu$ if $\mu(x) \leq \nu(x) \forall x \in X$ and we write $\mu \subset \nu$

if $\mu \subseteq \nu$ and there exists $x \in X$ such that $\mu(x) < \nu(x)$. Let $t \in [0, 1]$. We let $\mu_t = \{x \in X \mid \mu(x) \geq t\}$. Then μ_t is called the t-**cut** or the t-**level set** of μ.

We use the notation $\bigwedge$ to denote the minimum or infimum of a set of real numbers and $\bigvee$ to denote the maximum or supremum of a set of real numbers. Let $\mathscr{R}(X)$ denote the set of all relations on X. A fuzzy subset of $X \times X$ is called a **fuzzy binary relation** on X or simply a **fuzzy relation** on X. Let $\mathscr{F}\mathscr{R}(X)$ denote the set of all fuzzy relations on X. We let $\mathscr{F}\mathscr{R}^*(X) = \mathscr{F}\mathscr{P}^*(X \times X)$.

Definition 1.2 ($t - norm$) Let $\circledast$ be a fuzzy binary operation on $[0, 1]$. Then $\circledast$ is called a t-**norm** or fuzzy intersection if the following conditions hold: $\forall a, b, c \in [0, 1]$,

(1) $\circledast(a, 1) = a$ (boundary condition);
(2) $b \leq c$ implies $\circledast(a, b) \leq \circledast(a, c)$ (monotonicity);
(3) $\circledast(a, b) = \circledast(b, a)$ (commutativity);
(4) $\circledast(a, \circledast(b, c)) = \circledast(\circledast(a, b), c)$ (associativity).

Example 1.3 Some basic t-norms are

Standard intersection: $\circledast(a, b) = a \wedge b$;
Algebraic product: $\circledast(a, b) = ab$;
Bounded difference: $\circledast(a, b) = 0 \vee (a + b - 1)$;

Drastic intersection: $\circledast(a, b) = \begin{cases} a & \text{if } b = 1, \\ b & \text{if } a = 1, \\ 0 & \text{otherwise.} \end{cases}$

The bounded difference t-norm is also known as the Lukasiewicz t-norm.

Let $\circledast$ be a t-norm and let $a, b \in [0, 1]$. Then a and b are called **zero divisors** with respect to $\circledast$ if $a \neq 0, b \neq 0$, and $a \circledast b = 0$.

Let $\circledast$ denote a t-norm on $[0, 1]$. Let μ and ν be fuzzy subsets of X. Define the fuzzy subset $\mu \cap \nu$ of X by $\forall x \in X, (\mu \cap \nu)(x) = \mu(x) \circledast \nu(x)$.

Definition 1.4 (*Indifference Operator*) Let $\rho \in \mathscr{F}\mathscr{R}^*(X)$. Define the fuzzy relation ι on X by $\forall x, y \in X, \iota(x, y) = \rho(x, y) \wedge \rho(y, x)$. Then ι is called the **indifference operator** with respect to ρ.

Let $\rho \in \mathscr{F}\mathscr{R}^*(X)$ and let $*$ denote a arbitrary t-norm.

(1) If $\rho(x, x) = 1$ for all $x \in X$, then ρ is called **reflexive**.
(2) If $\rho(x, y) = \rho(y, x)$ for all $x, y \in X$, then ρ is called **symmetric**.
(3) If $\rho(x, y) > 0$ implies $\rho(y, x) = 0$ for all $x, y \in X$, then ρ is called **asymmetric**.
(4) If $\rho(x, y) * \rho(y, z) \leq \rho(x, z)$ for all $x, y, z \in X$, then ρ is called (max-$*$) **transitive**,
(5) If $\rho(x, y) > 0$ or $\rho(y, x) > 0$ for all $x, y \in X$, then ρ is called **complete**.
(6) If $\pi(x_1, x_2) * \pi(x_2, x_3) * \cdots * \pi(x_{n-1}, x_n) \leq \rho(x_1, x_n)$ for all $x_1, ..., x_n \in X$, then ρ is called **acyclic**.

Given $\rho \in \mathscr{F}\mathscr{R}(X)$, we often associate an asymmetric fuzzy binary relation π with ρ.

Definition 1.5 (*Fuzzy Binary Preference Relation*) Let $\rho \in \mathscr{F}\mathscr{R}(X)$. Define the fuzzy binary relations as follows: $\forall x, y \in X$,

$$\pi(x, y) = \begin{cases} \rho(x, y) & \text{if } \rho(x, y) > \rho(y, x) = 0 \\ 0 & \text{otherwise.} \end{cases} \tag{1.1}$$

In this book, unless otherwise specified, we assume that any asymmetric preference relation π associated with a preference relation ρ is the relation π defined by Eq. 1.1.

1.4 Fuzzy Maximal Sets

We conclude by considering the fuzzy maximal set. In later chapters, we present several public choice models that employ these sets to attempt to predict the outcome of the government formation.

Definition 1.6 Let $\mathscr{C}$ be a arbitrary function from $X \times X$ into $[0, 1]$. Define $M : \mathscr{F}\mathscr{R}^*(X) \times \mathscr{F}\mathscr{P}^*(X) \to \mathscr{F}\mathscr{P}^*(X)$ by $\forall(\rho, \mu) \in \mathscr{F}\mathscr{R}^*(X) \times \mathscr{F}\mathscr{P}^*(X)$,

$$M(\rho, \mu)(x) = \mu(x) * \underset{w \in \mu}{\circledast} \bigvee \{t \in [0, 1] \mid \mathscr{C}(x, w) * t \le \rho(x, w)\} \tag{1.2}$$

for all $x \in X$, where $*$ and $\circledast$ are (possibly identical) t–norms and $w \in \mu$ means that $\mu(w) > 0$, i,e, that $w \in \text{Supp}(\mu)$. Then $M(\rho, \mu)$ is called a **maximal fuzzy subset** associated with (ρ, μ).

In Georgescu (2007) the t–norm operator $\circledast$ is always the standard t–norm min, $\wedge$. We want our results to correspond to those in Georgescu, while having the freedom of letting $\circledast \equiv *$. Therefore, in Eq. 1.2 the t–norm $\circledast$ may or may not be different from $*$.

If $\mathscr{C}(x, w) = \mu(w) * \rho(w, x) \ \forall x, w \in X$ and $\circledast \equiv \wedge$, then M is the maximal fuzzy subset defined in Georgescu (2007). We denote this maximal fuzzy subset by M_G, but allow any t–norm $\circledast$ to replace $\wedge$. Whenever we use M_G, we assume $\circledast$ has no zero divisors.

$$M_G(\rho, \mu)(x) = \mu(x) * \underset{w \in \mu}{\circledast} \bigvee \{t \in [0, 1] \mid \mu(w) * \rho(w, x) * t \le \rho(x, w)\} \tag{1.3}$$

If $\mathscr{C}(x, w) = 1$ for $\forall x, w \in X$ and $\circledast \equiv * \equiv \wedge$, then M is the maximal fuzzy subset defined in Mordeson et al. (2008) (Definition 2.4, p. 311). We denote this maximal fuzzy subset by M_M.

$$M_M(\rho, \mu)(x) = \mu(x) \wedge \bigwedge_{w \in \mu} \bigvee \{t \in [0, 1] \mid \mu(x) \wedge \rho(x, w) \geq t\} \qquad (1.4)$$

Definition 1.6 produces a far nuanced fuzzy maximal set than the conventional maximal set. The conventional approach relies on a simple asymmetric relationship between two alternatives (x is strictly preferred to y). The fuzzy approach invites us to consider both μ, the degree to which each alternative is in the set of most preferred alternatives for each political actor, and π, the degree to which an alternative is preferred to another by each player. Moreover, while π is asymmetric, the fuzzy relation ρ goes one step further and addresses the symmetrical relationship between x and y. Thus, we are able to consider instances in which a political actor prefers x to y at some relatively high level, but that the same actor nonetheless might not consider y unacceptable. Hence x might be preferred to y at the set inclusion level .80, and y might be preferred to x at the .33 level. .33 would indicate that alternative y is less acceptable than x, but not altogether unacceptable.

Example 1.7 Let $X = \{x, y, z\}$. Let $\mu \in \mathscr{FP}^*(X)$ be such that $\mu(x) = 0$, $\mu(y) = 1/4$, and $\mu(z) = 1/2$. Let $\rho \in \mathscr{FR}^*(X)$ be such that $\rho(w, w) = 1$ for $\forall w \in X$, $\rho(x, y) = 1/4$, $\rho(y, z) = 1/8$, $\rho(x, z) = 3/4$, and $\rho(\_, \_) = 0$ otherwise. We proceed to calculate the maximal fuzzy subset M_G where $\circledast \equiv * \equiv \wedge$.

Since $\mu(x) = 0$, it is immediate that the maximal fuzzy subset membership grade for x is zero, $M_G(\rho, \mu)(x) = 0$.

Now the support of μ is the set $\{y, z\}$ so we proceed to calculate the maximal fuzzy subset membership grade for y.

$$
\begin{aligned}
M_G(\rho, \mu)(y) = \mu(y) * \Big[&\bigvee \{t \in [0, 1] \mid \mu(y) * \rho(y, y) * t \leq \rho(y, y)\} \\
&\circledast \bigvee \{t \in [0, 1] \mid \mu(z) * \rho(z, y) * t \leq \rho(y, z)\} \Big] \\
= &\; 1/4 * (1 \circledast 1) \\
&\; 1/4 \wedge (1 \wedge 1) \\
= &\; 1/4.
\end{aligned}
$$

Finally, we calculate the maximal fuzzy subset membership grade for z.

$$
\begin{aligned}
M_G(\rho, \mu)(z) = \mu(z) * \Big[&\bigvee \{t \in [0, 1] \mid \mu(y) * \rho(y, z) * t \leq \rho(z, y)\} \\
&\circledast \bigvee \{t \in [0, 1] \mid \mu(z) * \rho(z, z) * t \leq \rho(z, z)\} \Big] \\
= &\; 1/2 * (0 \circledast 1) \\
= &\; 1/2 \wedge (0 \wedge 1) \\
= &\; 0
\end{aligned}
$$

Proposition 1.8 *Let R be a relation on X and let S be a nonempty subset of X. Then*

$$M(1_R, 1_S)(x) = \begin{cases} 1 & \text{if } x \in S \text{ and } \forall w \in S, (x, w) \in R \\ 0 & \text{if } x \notin S \text{ or} \\ & \exists w \in X \text{ such that } w \in S, \mathscr{C}(w, x) \neq 0, (x, w) \notin R. \end{cases}$$

Proof We have that

$$\bigvee \{t \in [0, 1] \mid \mathscr{C}(x, w) * t \leq 1_R(x, w)\} = \begin{cases} 1 & \text{if } (x, w) \in R, \\ 1 & \text{if } \mathscr{C}(x, w) = 0, \\ 0 & \text{if } w \in S, C(w, x) \neq 0, (x, w) \notin R. \end{cases}$$

Proposition 1.8, we have that if $C(x, w) = 1_S(x) * \rho(w, x)$, then $M_G(\rho, \mu)(x) = 1$ if $x \in S$ and $\forall w \in S, [(x, w) \in R$ or $w \notin S$ or $(w, x) \notin R]$.

Proposition 1.9 *Let $(\rho, \mu) \in \mathscr{FR}^*(X) \times \mathscr{FP}^*(X)$ and $x \in X$. Then*

$$M_M(\rho, \mu)(x) = \mu(x) \wedge \bigwedge_{w \in \mu} \{\rho(x, w)\}.$$

Proof We assume that X is finite so that the supremum is the max and the maximum value of t in

$$\bigvee \{t \in [0, 1] \mid \mu(x) \wedge \rho(x, w) \geq t\}$$

is just $\mu(x) \wedge \rho(x, w)$. Thus we have that

$$M_M(\rho, \mu)(x) = \mu(x) \wedge \bigwedge_{w \in \mu} \bigvee \{t \in [0, 1] \mid \mu(x) \wedge \rho(x, w) \geq t\}$$

$$= \mu(x) \wedge \bigwedge_{w \in \mu} \{\mu(x) \wedge \rho(x, w)\}$$

$$= \mu(x) \wedge \bigwedge_{w \in \mu} \{\rho(x, y)\}.$$

Proposition 1.10 *Let $\circledast \equiv \wedge$. Then $\forall(\rho, \mu) \in \mathscr{FR}^*(X) \times \mathscr{FP}^*(X)$ we have that $M_M(\rho, \mu) \subseteq M_G(\rho, \mu)$.*

Proof Let $(\rho, \mu)\ \mathscr{FR}^*(X) \times \mathscr{FP}^*(X)$ and $x \in X$. Let $w_0 \in X$ be such that $\rho(x, w_0) = \bigwedge\{\rho(x, w) \mid w \in \text{Supp}(\mu)\}$ (X is finite). Then $M_M(\rho, \mu)(x) = \mu(x) \wedge w_0$. Now

$$M_G(\rho, \mu)(x) = \mu(x) * \bigwedge_{w \in \text{Supp}(\mu)} \bigvee \{t \in [0, 1] \mid \mu(w) * \rho(w, x) * t \leq \rho(x, w)\}$$

$$= \begin{cases} \mu(x) & \text{if } \mu(w) * \rho(w, x) \leq \rho(x, w), \ \forall w \in \text{Supp}(\mu), \\ \mu(x) * \bigwedge_{w \in \text{Supp}(\mu)} \{\rho(x, w) \mid \mu(w) * \rho(w, x) > \rho(x, w)\} & \text{otherwise.} \end{cases}$$

Let $w_1 = \bigwedge \{\rho(x, w) \mid \mu(w) * \rho(w, x) > \rho(x, w)$ and $w \in \text{Supp}(\mu)\}$ if $\exists w \in \text{Supp}(\mu)$ such that $\mu(w) * \rho(w, x) > \rho(x, w)$. Since $w_0 \leq w_1$, the desired result holds.

Proposition 1.11 Let $(\rho, \mu) \in \mathcal{F}\mathcal{R}^*(X) \times \mathcal{F}\mathcal{P}^*(X)$ and $x \in X$. Then

$$M(\rho, \mu)(x) = \begin{cases} \mu(x) & \text{if } \mathscr{C}(x, w) \leq \rho(x, w) \ \forall w \in \text{Supp}(\mu), \\ \mu(x) * t_x & \text{otherwise,} \end{cases}$$

where

$$t_x = \bigvee \{t \in [0, 1] \mid \mathscr{C}(x, w) * t \leq \rho(x, w)\}$$

if $\exists w \in \text{Supp}(\mu)$ such that $\mathscr{C}(x, w) > \rho(x, w)$.

In Proposition 1.11 ,

$$t_x = \bigwedge \{\rho(x, w) \mid \mathscr{C}(x, w) > \rho(x, w) \text{ and } w \in \text{Supp}(\mu)\}$$

if $* \equiv \wedge$ and

$$t_x = \bigwedge \{\rho(x, w)/\mathscr{C}(x, w) \mid \mathscr{C}(x, w) > \rho(x, w) \text{ and } w \in \text{Supp}(\mu)\}$$

if $*$ is the product operation (multiplication).

The definition of a fuzzy maximal subset associated with (ρ, μ) gives the degree to which the elements of X are maximal with respect to ρ and μ. The degree of maximality of an element x in X can neither be larger than the degree that it is a member of μ nor larger that the smallest degree for which x is preferred to those y which have positive membership in μ.

Theorem 1.12 Suppose $*$ has no zero divisors. Let $\rho \in \mathcal{F}\mathcal{R}^*(X)$ be reflexive and complete. Then $M(\rho, \mu) \neq \theta$ for all $\mu \in \mathcal{F}\mathcal{P}^*(X)$ if and only if ρ is partially acyclic.

Proof Suppose ρ is partially acyclic. Let $\mu \in \mathcal{F}\mathcal{P}^*(X)$. Then $\text{Supp}(\mu) \neq \emptyset$. Let $x_1 \in \text{Supp}(\mu)$. If $\rho(x_1, w) > 0 \forall w \in \text{Supp}(\mu)$, then $M(\rho, \mu)(x_1) > 0$ and so $M(\rho, \mu) \neq \theta$. Suppose there exists $x_2 \in \text{Supp}(\mu) \backslash \{x_1\}$ such that $\rho(x_1, x_2) = 0$. Then $\rho(x_2, x_1) > 0$ since ρ is complete. Thus $\pi(x_2, x_1) > 0$. Suppose there exists $x_1, \ldots, x_k \in \text{Supp}(\mu)$ such that $x_i \in \text{Supp}(\mu) \backslash \{x_1, \ldots, x_{i-1}\}$ and $\pi(x_i, x_{i-1}) > 0$ for $i = 2, \ldots, k$. If $\rho(x_k, w) > 0$ for all $w \in \text{Supp}(\mu)$, then $M(\rho, \mu) \neq \theta$ as above. Suppose this is not the case and there exists $x_{k+1} \in \text{Supp}(\mu) \backslash \{x_1, \ldots, x_k\}$ such that

$\rho(x_k, x_{k+1}) = 0$. Then $\rho(x_{k+1}, x_k) > 0$ since ρ is complete. Hence by induction, either there exists $x \in \text{Supp}(\mu)$ such that $\rho(x, w) > 0$ for all $w \in \text{Supp}(\mu)$, in which case $M(\rho, \mu) \neq \theta$, or since $\text{Supp}(\mu)$ is finite, $\text{Supp}(\mu) = \{x_1, \ldots, x_n\}$ is such that $\pi(x_i, x_{i-1}) > 0$ for $i = 2, \ldots, n$. Since ρ is partially acyclic and ρ is reflexive, $\rho(x_n, x_i) > 0$ for $i = 1, \ldots, n$. That is, $\rho(x_n, w) > 0$ for all $w \in \text{Supp}(\mu)$. Thus $M(\rho, \mu)(x_n) > 0$ and so $M(\rho, \mu) \neq \theta$.

Conversely, suppose $M(\rho, \mu) \neq \theta$ for all $\mu \in \mathscr{F}\mathscr{P}^*(X)$. Suppose $x_1, \ldots, x_n \in X$ are such that $\pi(x_1, x_2) > 0, \pi(x_2, x_3) > 0, \ldots, \pi(x_{n-1}, x_n) > 0$. We must show $\rho(x_1, x_n) > 0$. Let $S = \{x_1, \ldots, x_n\}$. Then $M(\rho, 1_S) \neq \theta$. Since $\pi(x_{i-1}, x_i) > 0$ and so $\rho(x_i, x_{i-1}) = 0, i = 2, \ldots, n$, we have that $M(\rho, 1_S)(x_i) = 0, i = 2, \ldots, n$. Thus it must be the case that $M(\rho, 1_S)(x_1) \neq 0$ and so $\rho(x_1, x_i) > 0$ for $i = 1, \ldots, n$. Thus $\rho(x_1, x_n) > 0$ and so ρ is partially acyclic.

References

Austen-Smith, D., Banks, J.: Positive Political Theory I. The University of Michigan Press, Ann Arbor (1999)

Benoit, K., Laver, M., Mikhalov, S.: Treating words as data with error: estimating uncertainty in the comparative manifesto measures. Am. J. Polit. Sci. **53**(2), 49–513 (2009)

Budge, I., Klingemann, H.D., Volkens, A., Bara, J., Tannenbaum, E. (eds.): Mapping Policy Preferences: Estimates for Parties, Elections, and Governments: 1945–1998. Oxford University Press, Oxford (2001)

Georgescu, I.: The similarity of fuzzy choice functions. Fuzzy Sets Syst. **158**, 1314–1326 (2007)

Klingemann, H.D., Volkens, A., Bara, J., Budge, I., McDonald, M.: Mapping Policy Prefences II: Estimates for Parties, Electors, and Governments in Eastern Europe, European Union, and OECD 1990–2003. Oxford University Press, Oxford (2006)

Laver, M. (ed.): Estimating the Policy Position of Political Actors. Routledge, London (2001)

Laver, M., Budge, I. (eds.): Party Policy and Goverment Coalitions. MacMillan Press, London (1992)

Mordeson, J.N., Bhutani, K., Clark, T.D.: The rationality of fuzzy choice functions. New Math. Nat. Comput. **4**, 309–327 (2008)

Chapter 2
Fuzzy Preferences: Extraction from Data and Their Use in Public Choice Models

Abstract We describe a method for extracting fuzzy preferences from the Comparative Manifesto Project (CMP) data that makes use of the bootstrap procedure designed by Benoit et al. (2009). We argue that fuzzy preferences are a better representation of the abstract concept of a player's preferences in public choice models. Instead of representing preferences as precise points, our fuzzy approach maps them as bounded areas in a subset of $\mathbb{R}^k$. In so doing, we eschew the conventional assumption that political actors have precise policy positions. Instead, fuzzy preferences permit us to conceive of actor's preferences as vague, but communicated accurately. We conclude the chapter by introducing our basic approach to using fuzzy preferences in fuzzy public choice models. We argue that a fuzzy public choice model satisfies some of the intuitive and practical problems faced by the conventional model. Moreover, a fuzzy public choice model allows us to shed the assumption that actors perceive shifts in utility in infinitely precise increments at the same granularity across and infinite policy space.

2.1 Introduction

A number of methods have been designed for the extraction of preference measures from empirical data. These measures represent players' preferences as precise "crisp" points in an infinite Euclidean space. As a consequence, the predictions made in these "crisp" models suffer under the assumptions of traditional $\mathbb{R}^k$ space. In contrast, fuzzy public choice models represent players' preferences as bounded areas in a subset of $\mathbb{R}^k$. Such models require a new method for extracting fuzzy preference measures from raw data.

This chapter presents a method for deriving fuzzy preference measures from raw text data. Rather than starting at square-one and collecting a new, raw dataset, we "fuzzify" an existing dataset, Comparative Manifesto Project (CMP) data. By doing so, we avoid the problems of selecting and collecting appropriate data. We may also

P. C. Casey et al., *Fuzzy Social Choice Models*, Studies in Fuzziness
and Soft Computing 318, DOI: 10.1007/978-3-319-08248-6_2,
© Springer International Publishing Switzerland 2014

be able to avoid some of the problems that have already been resolved in existing datasets.

We select CMP data because of its breadth and availability. It as also been applied broadly and shown to perform better when estimating the preferences of political actors in democratic systems outside of the United States. Moreover, the CMP offers an interesting opportunity to extract fuzzy preferences because it derives crisp preference measures from inherently fuzzy text data.

The CMP project involved human coding of text data in the form of manifestos published by political parties that detail the issues and policies the party promotes in an election year. It is in the stochastic process of writing the text, or text generation, that Beniot, Laver, and Mikhaylov (Benoit et al. 2009) find non-systematic (unbiased) error introduced into preference measures derived from text data. Language is inherently vague, and it is possible that, in the drafting and re-drafting of a political text, an author may not represent his preferences "precisely." Beniot, Laver, and Mikhaylov develop a bootstrapping method that replicates this stochastic process to calculate error scores for the CMP data. They find that much of the variation in parties' preferences over time can be accounted for by error, and that even the preferences of parties competing in the same country in the same election year may be statistically indistinguishable, making decision–making analysis based on CMP data impossible.

However, a fuzzy approach allows us to salvage the CMP data for application in decision-making analysis. Instead of conceiving actors' preferences as precise positions in policy space from which each incremental movement is less preferred, the fuzzy public choice model represents actors' preferences as bounded subsets of policy space over which players perceive shifts in utility. Instead of treating uncertainty as "noise" that eliminates the CMP's ability to identify precise policy positions, the fuzzy model uses uncertainty in a party's preferences to determine how the party perceives shifts in utility over a subset of policy space.

We utilize the bootstrapping method designed by Beniot, Laver, and Mikhaylov to derive fuzzy preference measures from the CMP data. This method projects a distribution of the possible outcomes of text generation. We then use this distribution of policy positions to determine the bounds of α-levels of the fuzzy numbers representing each party's preference profile. Finally, we apply these empirically-derived fuzzy preference measures in a public choice model. In later chapters, we further develop the basic model to permit predicting the outcome of the government formation process in parliamentary democracies.

2.2 Extracting Preference Measures from Empirical Data

There is a broad range of data and methods for collecting data available for the purpose of measuring the preferences of political players. All of these data and extraction methods have their benefits and disadvantages, but some data seem to be more amenable to the extraction of fuzzy preference measures than others. The most commonly used data for the purpose of measuring the preferences of political actors are roll-call votes, expert surveys, legislator surveys, and political texts.

The use of roll-call data to extract political actors' policy preferences has a number of advantages and disadvantages. Roll-call votes are the spoken "Yea or Nay" votes of individual legislators given on request for a piece of legislation. Data on roll-call votes is widespread and accessible. Processes such as NOMINATE (Poole and Rosenthal 2007) and Optimal Classification (Poole 2005) use roll-call data to locate players' ideal points in policy space. The more roll-call data available to such programs, the greater the precision of their estimates.

Despite its widespread use, a number of criticisms have been levied against the use of roll-call data to identify players' preferences. First of all, there are a number of institutional, political, and strategic factors that may influence the way a player votes. Votes may not reflect players' sincere preferences, but may reflect their preferences as they are translated by the institutional and political constraints within which they are voting. There is a further causal problem inherent in using roll-call votes as inputs to describe the behavior of actors in political situations when it is the same political context from which the vote was a behavioral output (Laver 2001, p. 239).

Furthermore, roll-call data may not be amenable to the extraction of fuzzy preference measures. Roll-call data is based on a system of "Yea," "Nay," or "Abstain." The values "Yea" or "Nay" are dichotomous and represent crisp positions comparable to those used in traditional set theory from which Euclidean models are derived. A vote "Yea" puts a policy fully within the set of a player's preferences (i.e., $\alpha = 1$) and a vote "Nay" puts a policy strictly out of a player's preferences (i.e., $\alpha = 0$). While such data can be used to identify the core and the support of a fuzzy number, the ability to identify other discrete α-levels is limited.

Nevertheless, Potter (2007) extracts fuzzy measures of individual legislators' preferences from roll-call data. Potter assigns a vote 'Abstain' to an α-level of $\alpha = 0.5$, assuming that abstaining from a vote is the same as being indifferent. However, such an assumption is problematic in that the choice to abstain from a vote can be a sign of disapproval, and even an attempt to undermine the quorum necessary to hold a vote. When applied to NOMINATE and OC, a vote "Abstain" is coded as missing data. Without the ability to define α-levels other than $\alpha = 1$ and $\alpha = 0$, much of the power and nuance of the fuzzy model is lost.

A better alternative may be to derive preference information from expert or legislator surveys. Experts can respond to survey questions that provide an intensity scale that can easily be applied to the discrete α-levels derived from players' fuzzy preferences. However, expert surveys are subject to the same criticism as roll-call data in that experts' judgments are derived from the same actions they are seeking to explain. Beyond this potential problem of tautology, expert survey data can be expensive and difficult to gather, making its accessibility limited.

Furthermore, survey data is not replicable. Once a survey has been taken at a specific time t_1, it cannot be taken at that time again. Instead any survey taken at a later time t_2, has to be understood to be different and informed differently than the original survey. This is of a particular problem if we decide to expand our set of questions. For example, consider if we would like to add a new policy dimension

"Immigration" to our models of political decision-making. Without past questions about immigration policy, we cannot hope to compare surveys at t_1 and t_2. Since it is impossible to go back and add questions about immigration to the survey at t_1, we are not able to compare the policy positions of actors at these two different times. Due to the constraints upon gathering survey data and the continuing problem of tautology, expert data is less than ideal.

Alternatively, surveys of actual political actors allow us to capture the same kinds of scaling as expert data and avoid the problem of causal tautology. By surveying political actors directly, scholars can collect data about their preferences does not result from actions within a specific political context. However, legislator surveys are subject to the same problems of resource-intensiveness, time-sensitivity, and lack of replicability as expert surveys, and therefore are impractical for the extraction of preference measures.

Political texts are a good alternative to the other forms of data used for measuring players' preferences. Political texts published by political actors are widely available and accessible in most developed democracies, and can be obtained with limited resources. The causal tautology is escaped, because the inputs for the prediction of players' behavior are no longer the outputs of their behavior in the same political context. Also, the policy emphases of political parties in political texts tells us which issues they consider salient, giving us a sense of what issues we should consider plausible dimensions at that time. Therefore, we know what issue dimensions to consider, or can infer a player's indifference to an issue dimension, without having to turn to direct questions about the issue. This releases us from the time-sensitivity and replicability problem of survey data. Finally, differing degrees of intensity of preferences can be derived based on the policy emphases in the language of the texts, making it particularly amenable to the measurement of fuzzy preferences.

2.2.1 Extracting Preferences from the CMP Data

Among the most widely used data sets of political texts from which preferences measures have been derived is the Comparative Manifesto Project (CMP). The CMP project was undertaken by a research group concerned with locating the policy positions of political parties based on the policy emphases within the parties' electoral manifestos. An electoral manifesto is a document released by a political party prior to an election that contains a wide range of authoritative and representative statements about the policy positions of the political party as a whole (Budge et al. 2001; Klingemann et al. 2006). The CMP represents its cases as country-party-year. All salient parties from every free democratic election from 1945 to 2005 are included. A salient party is any political party that is either likely to be a member of a governing coalition or that can affect the outcome of a governing coalition. Thus, each case represents a salient political party in a country each year it competes in a free democratic election. The CMP dataset includes 780 political parties from 54 countries in 529 election years for a total of 3,108 cases (country-party-years).

Each party manifesto in the CMP database is parsed into "quasi-sentences"—or phrases that reflect complete thoughts—by human coders and placed numerically into one of 56 policy dimensions, or designated as "unknown" in the case of insufficient data (Budge et al. 2001; Klingemann et al. 2006). Insufficient or missing data are addressed by collecting and analyzing supplemental party materials. When election manifestos released by the political parties themselves are unavailable, the CMP seeks published documents fulfilling the above criteria. The CMP collects data only for significant political parties which are identified based on their potential for being members of a governing coalition or their ability to affect the tactics of parties competing for a position in the government.

Quasi-sentences are the basic unit of meaning identified and coded by the CMP. They can be understood as "complete thoughts" either in the form of sentences or sentence fragments (Budge et al. 2001; Klingemann et al. 2006). CMP coders placed each quasi-sentence on the policy dimension on which it best fits. The sum of the quasi-sentences assigned to each party is that party's score on that dimension.

Subsets of the 56 dimensions coded in the CMP database, or *riles*, can be created to compare parties on any number of political concepts. Among the most popular of such *riles* is a single-dimensional left-right political scale developed by Laver and Budge (1992, pp. 23–30). They constructed the left-right scalar using exploratory factor analyses to determine whether issues that are theoretically related share a common dimension empirically. They then construct the *rile* by adding scores on common dimensions. The left-wing policy categories include an emphasis on democracy, state intervention, and peace and cooperation as well as a positive orientation toward social services, state education, and labor groups.[1] The right-wing policy categories include and emphasis on freedom, domestic human rights, the military, capitalist economics, and social conservatism.[2] The position of political parties on the *rile* dimension is determined by the sum of the proportion of the manifesto devoted to right-wing policy references minus the sum of the proportion devoted to left-wing references.

A number of criticisms have been levied against the process employed by the CMP. The strongest criticism of the CMP is the problem of inter-coder reliability. The inter-coder reliability problem refers to the dilemma that two different coders or

---

[1] State intervention and Peace and cooperation are additive dimensions established by exploratory factor analyses of theoretically similar CMP policy categories. State intervention includes the original CMP categories Regulation of capitalism (PER403), Economic planning (PER404), Protectionism: positive (PER406), Controlled economy (PER412), and Nationalization (PER413). Peace and cooperation is constructed from the categories Decolonization (PER103), Military: negative (PER105), Peace (PER106), and Internationalism: positive (PER107).

[2] Capitalist economics and Social conservatism are additive dimensions established by exploratory factor analyses of theoretically similar CMP policy categories. Capitalist economics includes the CMP categories Free enterprise (PER401), Incentives (PER402), Protectionism: negative (PER407), Economic orthodoxy and efficiency (PER414), and Social services expansion: negative (PER 505). Social conservatism includes the categories Constitutionalism: positive (PER203), Government effectiveness and authority (PER305), National way of life: positive (PER601), Traditional morality: positive (PER603), Law and order (PER 605), and National effort, social harmony (PER606).

the same coder at two different times may code the same text differently (Mikhaylov et al. 2012). Error resulting from the identification and coding of quasi-sentences reflects uncertainty about the accuracy of the coders' reading of the text, but it does not tell us anything about the political actor's preferences. Moreover, it is specific to the CMP coding process.

Benoit, Laver, and Mikhaylov (BLM) (2009) argue that error may result as well from published texts failing to communicate the preferences of the party accurately. This type of error, error introduced by the process of authorship or "text generation" (Benoit et al. 2009), reflects uncertainty about the relationship between the author's intended message and communicated message and is problematic for any process extracting preference measures from text data. In order to get at this problem, BLM develop a method to measure uncertainty related to this type of error.

BLM (2009) describe text generation as a stochastic process by which an author attempts to communicate his policy position by writing a text. The authors start with an intended message μ that he seeks to communicate. BLM assume that the authors intended message μ is the author's actual policy position, and that the author is not dissembling. We can imagine that the process of authorship involves a number of different versions of any text, including intended drafts in the mind of the author, drafts written and stored by the author, written and thrown away, and the actual text that is published coded by the CMP. Each one of these texts communicates a policy position that is an approximation of the author's intended message μ. The process of communicating the intended message μ with some text τ is stochastic and therefore allows for random error to enter into the information communicated in τ. Through the stochastic process of text generation, authors may accidentally misallocate quasi-sentences to the different policy categories they mention, thus overemphasizing some policies and underemphasizing others. This results in the author communicating a policy position not in-line with their intended message. Therefore, the policy position communicated by the text may be another policy position on the relevant dimension with some random error differentiating it from the author's actual position. By this logic, any attempt to derive the true preferences of a political actor from a political text will be affected by some random error.

BLM use a bootstrapping procedure to derive uncertainty scores based on the error in text data-derived preference measures caused by the stochastic process of text generation. Salience theory, the theoretical grounding of the CMP, assumes that authors emphasize certain policies by making reference to them.[3] Authors only mention policies they support, and the more an author mentions a policy (i.e., the more quasi-sentences the author commits to the policy), the more strongly the author prefers it. The bootstrap procedure simulates the stochastic text generation process by taking the number of quasi-sentences in a document and the categories to which those quasi-sentences are assigned and then randomly reassigning the quasi-sentences to the set of categories. This process is repeated 1,000 times to produce a standard error of policy uncertainty, which is assumed to represent the error, or noise, in the CMP data.

---

[3] Budge et al. (2001), Klingemann et al. (2006).

As an illustration of the bootstrap procedure, assume a document of 20 quasi-sentences assigned to four different categories from the CMP; we will call these categories a, b, c, and d. Assume that the quasi-sentences in the text are distributed equally across the four issue categories, such that five quasi-sentences are devoted to each category. The bootstrap procedure would then hold the number of quasi-sentences in the document constant at 20, but reassign them randomly to each of the four issue categories. The bootstrap does not assign quasi-sentences to any issue category not included in the original coding, assuming that, by not mentioning an issue category, a party is explicitly not showing support for that policy. So an example of a bootstrap reiteration of the text in this case would be the reassignment of the quasi-sentences from $a = 5$, $b = 5$, $c = 5$, and $d = 5$ to $a = 6$, $b = 4$, $c = 3$, and $d = 7$. The bootstrap procedure simulates the stochastic process of text generation and reiterates it 1,000 times. The procedure then calculates the standard error from the points projected from the set of iterations to derive an uncertainty score for the data.

BLM conclude that much of the perceived change in parties' ideal policy positions over time calculated by the CMP can actually be accounted for by unbiased error introduced by the stochastic text generation process. In fact, BLM find that observed changes in the parties' policy positions are statistically significant in only 28.1% (less than one-third) of the cases they investigate.[4] Thus, BLM conclude that much of the observed changes in parties' policy positions over time in the CMP data is "noise." Similarly, this causes problems for comparing parties' policy positions in a specific country and election year in order to make inferences about political outcomes such as government formation and legislation. BLM demonstrate that, due to unbiased error introduced by the stochastic text generation process, many parties' policy positions are no longer statistically distinguishable. Because Euclidean public choice models are based on differences between actors' policy positions measured in distance, such error renders the use of CMP scores to determine decision–making outcomes impossible. In fact, even if parties' policy positions are still statistically distinguishable, the inability to determine the precise position of each parties' ideal point in policy space makes determining the distance between parties' ideal points impossible. Thus, the unbiased error introduced into the CMP data by the stochastic text generation process that is calculated by BLM demonstrates that the CMP data is unusable for Euclidean public choice models.

The bootstrap procedure created by BLM (2009) is a major contribution to studies utilizing not only CMP data but all methods of extracting preference information from texts. However, their analysis of how error is introduced into preference data by the stochastic text generation process relies on some fundamental assumptions about how humans think and communicate. First, BLM assume that the text author's intended message (and, by assumption, his actual policy position) is a single precise point in policy space. Second, they assume that each text communicates a single

---

[4] BLM only consider cases without the extended categories add in the second-generation CMP data (Klingemann et al. 2006) and do not consider cases in which 99.99% of quasi-sentences are uncoded (Sweden from 1948 to 1982 and all Norway).

precise point in the policy space. By these assumptions, it would be possible for an author to communicate his intended message μ in a text τ if he used the right language. However, BLM assume that the text generation process is stochastic and thereby introduces error into the communication of preferences.

Moreover, Benoit et al.'s bootstrapping procedure creates a significant problem for the conventional public choice models. If, as they assume, the error derived from the bootstrapping represents the writer's inability to adequately communicate their precise policy position in Euclidean space, then the further assumption is that a party's policy position does not change between election years. Instead, any change in quasi-sentence allocation is error. However, without variability in party policy positions, the CMP data becomes less than ideal for estimating preferences as the basis upon which outcomes will be predicted by public choice models.

2.2.2 A Method for Extracting Fuzzy Preference Profiles from CMP Data

The assumptions of BLM are essentially Euclidean in nature. In the Euclidean tradition, BLM conceive of each author's real policy position, intended message, and policy position communicated by the author's text as precise points in policy space. Therefore, error introduced into the CMP data by the stochastic text generation process comes from a statistically significant difference between the ideal point position of the author's intended message and the text's message.

A fuzzy approach to estimating preferences would permit us to conceive of the uncertainty caused by the text generation process not as error between the text's message and the author's intended message, but rather as uncertainty within the author's actual preferences. Thus, rather than treat actor's policy positions as precise points in policy space as in Euclidean models, uncertainty in actors' preferences can be represented as subsets of policy space over which the actors perceive shifts in utility. In this way, vagueness about the "precise" position of an actor's ideal point is not problematic, and in fact is information about the actor's preferences and the likelihood that the actor will agree with other actors. In essence, a fuzzy approach invites us to interpret the stochastic text generation process other than authors erring in their emphasis of certain policy positions. Rather, we see them as flexible in the priority they place upon certain policies.

The bootstrap method designed by BLM estimates error in the CMP data by simulating the stochastic process of text generation for each case in 1,000 reiterations. Each simulation predicts a point meant to represent a possible policy position communicated in a text following the text generation process. This process produces a distribution of points around the point estimated by the CMP (treated as the median of the distribution). From this distribution, a density function can be derived from which BLM estimate error in the CMP data introduced by the stochastic text generation process (depicted in Fig. 2.1).

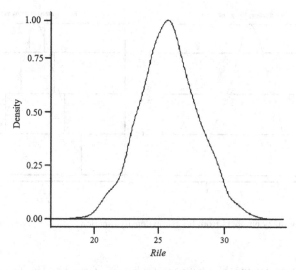

Fig. 2.1 Density function derived from bootstrapping procedure

In Fig. 2.1, there are two axes. The x-axis is the *rile* issue dimension upon which the bootstrap procedure estimates the placement of policy positions communicated by the stochastic text generation process. The *rile* dimension (x-axis) is used for the categories of random distribution of quasi-sentences per manifesto. The y axis represents the density distribution of quasi-sentences after 1,000 iterations. The y-axis is the density of the predicted policy positions at a certain point on the x-axis. The density function is single-peaked and resembles a normal curve.

We use the density function resulting from BLM's bootstrap procedure to derive a fuzzy preference. Again, we place the density function in two-dimensional space. We use the y-axis of the function, the density of the points projected on the x-axis, as a surrogate for the party's utility if policy is located on that position of the issue dimension. Therefore, we assume the denser the distribution of points at a position, the more that position is preferred by a party. We identify the highest density position as the set of policies perfectly in the set of the party's preferences with an α-level of $\alpha = 1.00$. The boundary of the density function, where distribution of points ends, we identify as the boundary of the fuzzy number, beyond which the party's utility is $\alpha = 0.00$. Fuzzy preference profiles are created with the introduction of cut points at density levels 0.25, 0.5, and 0.75 to begin creating the alpha levels of indifference. Once the alpha cuts are made to the density functions, we further adjust the preference profile to represent areas of indifference in policy space. This is done by reshaping the density function between alpha cuts into rectangles with the same area as the original curve.

Figure 2.2 illustrates the process. Between $\alpha = 1.00$ and $\alpha = 0.00$ we make four equidistant cuts (α-cuts) perpendicular to the y-axis; $\alpha = 0.75$, $\alpha = 0.50$, and $\alpha = 0.25$ (depicted in Fig. 2.2a). The area and center of gravity for the segment

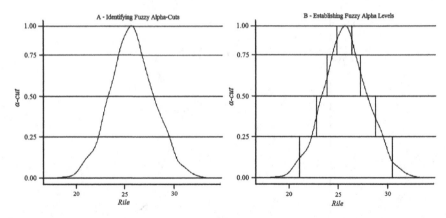

Fig. 2.2 Deriving fuzzy numbers from bootstrap density function

of the distribution sandwiched between each pair of α-cuts. The function is then re-shaped into a rectangular shape with the same area and positioned at the same center of gravity (see Fig. 2.2b). This process allows us to shape the density function into a discrete fuzzy number with minimal loss of information. Each party's fuzzy preference profile was then placed on a *rile* dimension with other parties from its country and the year it competed in the election.

2.3 The Conventional Public Choice Model

Now that we have a means to extract fuzzy preferences from CMP data, we are in a position to employ them in public choice models to see if we can improve our predictions. We conclude this chapter by introducing the book's approach to using fuzzy preferences in public choice models. We begin by giving considering the conventional public choice model. We then juxtapose our fuzzy public choice model against the conventional approach.

Traditional Euclidean public choice models are based on the assumption that there exists a universal set of policies X, such that, for all policies x and y in X, any player i has a binary preference relation R over these policies Austen-Smith and Banks (1999, pp. 1–3). In other words, X is a set of all conceivable policies and x and y are policies.

Any political player perceives x and y and has some preference relation R between them, for example xRy. The statement xRy means that "x is at least as good as y." If, for some player i, xRy and yRx we say that "x and y are at least as good as each other" or "i is indifferent between x and y" and we use the binary relation I to write the statement xIy. If, instead, xRy and not yRx then we say "x is strictly preferred to y" and we use the binary relation P to write the statement xPy (Austen-Smith and Banks 1999, p. 2).

Implicit in the statement we just made is the assumption of completeness: that for all policies x and y in X, a player i holds some preference relation over them such that either xRy or yRx or both (p. 3). In addition, the traditional model necessitates the assumption of reflexivity, xRx, that x is at least as good as itself.

When developing a public choice model, we are concerned with the preferences of individual players only in so far as their preferences determine the group preference. When seeking a group preference we are interested in the existence of a maximal set. Given some decision-rule, the maximal set is the set of all policies x in X such that, for all policies y in X, x is at least as good as y, or xRy. In other words, the maximal set is the set of all policies for which there is no other policy that the group strictly prefers.

The existence of a maximal set requires that the group preference be complete and reflexive. In much of the literature on models of public choice, transitivity is considered a requirement for a maximal set to exist. When we assume rational individuals, we assume that their preferences are transitive. That means, if player i prefers x to y (xRy) and y to z (yRz), then i must also prefer x to z (xRz), for a complete preference profile of $xRyRz$.

However, the rational choice assumption applies to individuals, not groups (see Arrow 1951). Therefore, we cannot assume that the group's preferences are transitive. This leaves the possibility that a group's preferences may cycle. Cycling occurs when the group's preferences are intransitive; for example, the group prefers x to y and y to z, but prefers z to x; or the group's preferences are $xRyRzRx$. In such a case, there is no element in the maximal set; the maximal set is empty. Austen-Smith and Banks (1999, pp. 4–6) demonstrate that transitivity is a sufficient but not necessary requirement for a maximal set to exist. A weaker assumption, that group preferences are acyclic, is sufficient for a maximal set to exist. Acyclicity assumes that for all policies x, y, z, w in X, if $xPyPzPw$ then xRw. If a group's preferences are complete, reflexive, and acyclic, then the maximal set is said to be nonempty.

While acyclicity is a minimal requirement for a nonempty maximal set, there is nothing in the definition of group preference to guarantee that they will be acyclic. To guarantee acyclic preferences, the Euclidean model assumes that the preferences of individuals in a group are single-peaked and monotonic. If a player's preferences are single-peaked and monotonic, this means that there is some point in the policy space that the player holds as ideal and that every incremental movement away from this point in the policy space is strictly less preferred than any point closer to the player's ideal. This is also called the Euclidean distance assumption.

The Euclidean distance assumption is based on the idea that players view preferences as if they existed in an n-dimensional space, where n is any number of dimensions. The simplest Euclidean models feature two dimensions in which the x-axis is a continuum of possible policies and the y-axis is the player's utility at that given policy. In this book, we follow the convention in political science of labeling these models single-dimensional because only one *policy* dimension is depicted. Figure 2.3 depicts a single-dimensional model in which the x-axis is some policy and the y-axis is the player's utility. The single-peaked preference profile for a single player is depicted in the model. Along the x-axis are a number of points in the

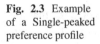

Fig. 2.3 Example
of a Single-peaked
preference profile

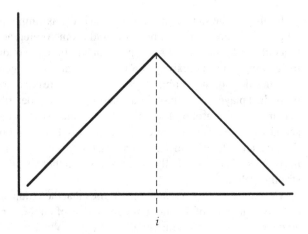

policy space. Point i is the player's ideal point, and is where the player's utility
peaks. Notice that as the policy points move away from the player's ideal point i the
player's utility decreases, i.e., they are less preferred by the player.

Figure 2.3 is an example of a *single-peaked preference profile*. This term implies
that as policy alternatives move away from a player's ideal point, the player's perceived utility decreases monotonically. Traditional Euclidean models assume that all
players' preference profiles are single-peaked.

Another way to demonstrate a player's preferences over a policy is by drawing the
player's circular indifference curve. Assuming the player's preferences are single-
peaked, we know that each incremental move from the player's ideal point is less
preferred. If we measure the distance between the player's ideal point and a policy
point x, and then draw a circle that captures all points y equidistant from the player's
ideal point, then all points y on the circle are indifferent to x (xRy and yRx, therefore
xIy). Any point bound by the circle is preferred to the points on the circle, this is
called the winset of x. Any point beyond the bounds of the circle are less preferred
than the points on the circle and are referred to as the inverse winset of x. In a single-
dimensional model, there are only two policy points on the indifference curve where
the circle intersects the dimension (Fig. 2.4). However, as the number of dimensions
expands to two or greater, the number of points on the indifference curve becomes
infinite.

In the single-dimensional Euclidean model, given that players all have single-
peaked preferences and decide by a simple majority voting rule, there is always a
nonempty maximal set; i.e., there exists some policy point x such that for all other
policy points y, the group holds x to be at least as good as y, or y is not strictly
preferred to x, or xRy. This is demonstrated by what is called Black's Median
Voter Theorem (Black 1958), which states that, given an odd number of players
with single-peaked preferences along a single dimension, the group outcome will
always be the ideal point of the median voter. This can be seen in Fig. 2.5 below.
Figure 2.5 represents three players each with a single-peaked preference profile; such

Fig. 2.4 Circular indifference

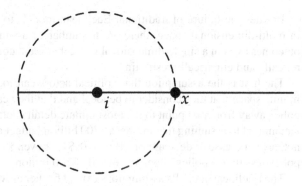

Fig. 2.5 Black's median voter theorem

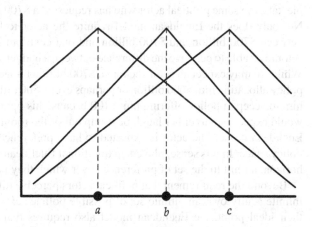

that for Player 1 aR_1bR_1c, for Player 2 bR_2cR_2a, and for Player 3 cR_3bR_3a based on Euclidean distance. Therefore, for either Player 1 or Player 3, Player 2's ideal point b is preferable to the ideal point of the further player. In a tournament, either Player 1 or Player 3 would always agree to Player 2's ideal point b rather than accept their least preferred option.

Once the number of dimensions in the Euclidean model exceeds one, the outcomes of a simple majority game become less certain. In fact, except under very specific conditions,[5] the set of possible outcomes of a simple majority game can be the entire Euclidean space (Austen-Smith and Banks 1999, p. 180). This is often referred to as McKelvey's Chaos Theorem. This is the result of cycling throughout the Euclidean space, and predicts unstable outcomes as the result of simple majority games in which there exist more than two-issue dimensions.

---

[5] Under conditions of perfect symmetry in two dimensional space, called the Plott conditions, the maximal set is nonempty (Austen-Smith and Banks 1999, pp. 142–149).

Besides the failure of traditional Euclidean models to produce stable outcomes in multi-dimensional space, there are a number of assumptions integral to stable outcomes even in a single-dimensional game that seem counterintuitive, if not theoretically and empirically restricting.

The first is the assumption that political actors can locate a precise point in an infinite space that they consider to be ideal and that they can distinguish any shift in policy away from that point to the most minute detail (Potter 2007, pp. 8–9). This is tantamount to assuming that in, say, a $700 billion budget allocation, a player would notice an increase or decrease of $100, $10, $1, or even $0.01, and would prefer that policy less than a policy allocating exactly $700 billion.

The Euclidean model's assumptions extend further as we consider that any player must also hold preferences over the entire infinite policy space (pp. 9–10). For example, take the same political actor who has requested a $700 billion budget allocation. Not only does the Euclidean model require the actor to hold distinct preferences between $700 billion and $700 billion and one cent, but it also requires the same actor to be able to perceive the difference between a budget allocation of $10 and $11. While we may expect an actor seeking a $700 billion budget allocation to consider a policy allocating him $600 billion or perhaps even $500 billion, we would not expect him to accept a policy offering him $100 because his options were that or $90. We would expect him to either laugh or stomp out of the room furious. However, in the Euclidean model, the actor is expected to have preferences over the entire infinite policy space. In this sense, players in traditional Euclidean models lack a preference horizon, a limit to the set of preferences over which they perceive shifts in utility.

Beyond the requirement that political actors perceive differences in policies with infinite acuity over an infinite set of possible policies extending infinitely far from their ideal point, the Euclidean model also requires that political actors evaluate policies using the same metric or granularity (pp. 10–11). In the Euclidean model, all players are assumed to perceive shifts in policy at the same intensity, with no account for possible differences in the actors' perception of policy space. Returning to our $700 billion budget allocation, the player requesting the budget allocation, we assume, has a deep understanding of the requirements of his program, and therefore understands the great necessity of the money he has requested. While he may be able to tolerate some minor shifts in the amount allocated to his program, say a few hundred thousand or a few million dollars, major shifts from his budget request would be significantly less preferred by him, perhaps even lie outside his preference horizon. Alternatively, another player in the budget allocation process who only sees the money as tax-payer dollars may not sense a big difference between $700 billion and $200 billion. What seems to be marginally as satisfying a policy to the latter player would be disastrous in the eyes of the former. Clearly, these players would view policy shifts in different granularities, but the Euclidean model cannot account for this.

2.4 A Fuzzy Public Choice Model

The application of fuzzy set theory to public choice models allows us to escape many of the pitfalls of traditional Euclidean models. By allowing for indifference in large areas of players' preferences, the fuzzy model avoids the strong assumptions of the traditional Euclidean model of public choice that players perceive shifts in policy with infinite acuity. Rather, the fuzzy model proposes that there are large areas of policy in which a player will experience no change in preference.

In reference to our previous example, Fig. 2.6 depicts the fuzzy preference profile of our player requesting a budget allocation of $700 billion dollars may actually be happy with anything in a range from $650 to $800 billion dollars. Anything less than $650 billion or more than $800 billion would then be less satisfactory for him. Let us say that our player's preference for a budget allocation decreases as it moves from $650 billion to $400 billion and below $400 billion our player is completely dissatisfied. Similarly, our player's satisfaction decreases as the budget allocation exceeds $800 billion and he is completely dissatisfied if the budget allocation exceeds $950 billion, seeing the allocation as exorbitant and wasteful.

We then refer to the range of $650 billion to $800 billion as the core of the fuzzy number representing our player's preferences at which the membership value of a policy in our player's set of preferences equals 1. In Fig. 2.6, the core of the fuzzy preference profile is the flat top of the trapezoidal fuzzy number. Notice that the crisp point of the Euclidean model has been replaced by a flat area of indifference, where the player perceives no difference between the policies. The support of the fuzzy number representing the player's preference profile is then the base of the fuzzy number where the membership value of a policy in the player's preferences is greater than zero. The continuous lines from the core of the trapezoidal fuzzy number to the base of the support are then the monotonically decreasing preferences of the player. Like in crisp, player's preferences decrease incrementally along these continuous slopes.

As can be seen in Fig. 2.6, we can draw lines through the slopes that descend between the core and support of the fuzzy preference profile that we refer to as alpha-cuts. An alpha-cut, or α-cut, selects those elements whose membership values are at least equal to α. A strong α-cut selects only those elements whose membership values are strictly greater than α. The α-cut at one, or $\alpha = 1$, is the core and the strong α-cut at zero is the support.

The core is always a subset of the support. That means that the continuous lines running from the core to the support always decrease monotonically. However, as can be seen in Fig. 2.6, the rate of decrease on either side of the fuzzy number need not occur at the same rate. The fuzzy model does not measure player's preferences by Euclidean distance but by the shape of the fuzzy number representing the player's preferences. The fuzzy model thereby abandons the Euclidean distance assumption, and we no longer need to assume that player's perceive shifts in policy at the same metric. Furthermore, the fuzzy set A is a subset of the set of alternatives X, therefore the support of the fuzzy number is necessarily a subset of the policy space. By

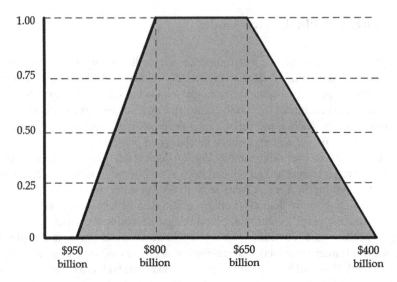

Fig. 2.6 Continuous fuzzy preference profile

Fig. 2.7 Circular indifference

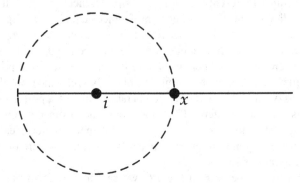

representing player's preferences as a fuzzy number, we define a player's preference horizon; a point beyond which the player derives no utility.

While the fuzzy model alleviates these two problems associated with the Euclidean assumptions of traditional models, the continuous lines decreasing monotonically from the core to the support retains the problem that player's are expected to make infinitely fine judgments between policies. However, we can alleviate this problem by representing players' preferences as discrete fuzzy numbers rather than continuous fuzzy numbers, as depicted in Fig. 2.7.

In Fig. 2.8, the preference profile of our player requesting a budget allocation of $700 billion is represented as a discrete fuzzy number. The player's preferences are now represented by a set of stacked boxes rather than a smooth curve. Each block

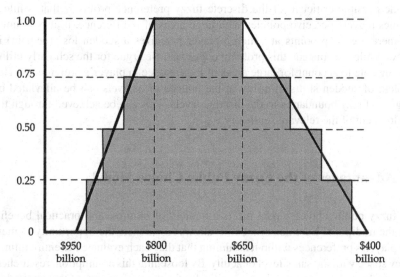

Fig. 2.8 Tiered fuzzy preference profile

represents an α-cut. The difference between representing a player's preferences as a discrete fuzzy number and representing them as a continuous fuzzy number is granularity. A continuous fuzzy number assumes that the player can perceive shifts in policy along the slope descending from the core to the support with infinite acuity. A discrete fuzzy number assumes that the player has areas of indifference at which he cannot determine the difference between policies in terms of his utility. Where a player formerly perceived shifts in utility from 0.84 to 0.8397 to 0.80 to 0.76, he now groups these together and sees them all at a single-level; $\alpha = 0.75$. As can be seen in Fig. 2.8, $\mu A = 0.75$ if $0.75 \leq \mu B < 1.00$. So, while our player may perceive the difference between \$625 and \$600 billion conceptually, the difference means nothing to him in terms of utility, as his utility for both is $\alpha = 0.75$.

We can understand the discrete α-cuts of any element's inclusion in the set of a player's preferences. We already know that at $\alpha = 1.00$ we say that a policy is perfectly in the set of a player's preferences, or is ideal. If $\alpha = 0.00$ for a certain policy, the policy is completely out of the player's set of preferences, or completely not preferred by the player. Between these two extremes of ideal preference and complete rejection are areas of ambiguous preference. At $\alpha = 0.75$ we may say that a policy is more in the set of a player's preferences than out, or the player prefers the policy more than does not. At $\alpha = 0.25$ we may say that a policy is more out of the set of a player's preferences than in the set, and therefore the player more dislikes than likes the policy. At $\alpha = 0.5$ a policy is neither in nor out of the set of a player's preferences, and therefore we can say his preferences over these policies is ambiguous.

One possible criticism of the discrete fuzzy preference profile is that, while it assumes players perceive policies with large areas of indifference, it also assumes that there are crisp points at which a player perceives a sudden loss (or gain) in utility. While we concede this problem does exist, we argue for the scholarly utility of being able to account for large areas of indifference in player's preferences. This problem of sudden shifts in utility at the bounds of α-levels can be alleviated by assigning fuzzy boundaries to the discrete levels. This can be achieved through the development of the relevant set theory.

2.5 Advantages of the Fuzzy Public Choice Model

The fuzzy public choice model offers a number of intuitive and practical benefits over the traditional Euclidean model. It allows researchers the opportunity to measure player's preferences without assuming that they each evaluate the entire infinite policy space with the same level of acuity. By loosening this assumption, researchers can actually tailor fuzzy numbers specifically for the preferences of the actors they are studying. Furthermore, they can find areas in which no compromise is possible between actors; where their preference horizons end before they overlap. Most importantly, however, researchers can use the fuzzy model to make stable predictions about group choice without importing further assumptions such as institutional procedures or sophisticated voting.

However, the most important advantage of a fuzzy public choice model is that it predicts stable outcomes more consistently than does the traditional Euclidean model. We will demonstrate the ability of several fuzzy public choice models to predict government formation outcomes using the CMP empirical data. In Chap. 4, we consider two single-dimensional fuzzy public choice models. The first makes use of the fuzzy maximal set, and the second will make use of the fuzzy Pareto set to make predictions. In Chap. 6, we focus on the ability of a fuzzy two-dimensional model to predict the same government formation outcomes.

All of the fuzzy public choice models that we present begin by using fuzzy preferences that conceive of preferences as subsets of policy space over which the actor assigns different utility values. Benoit et al. assume that a party preference profile contains a large amount of error and a potential overlap in preferences does not equal a willingness to agree between parties. Our approach argues that overlaps, or intersections, between actors' preference profiles represent subsets of policy over which actors can agree. Moreover, our fuzzy public choice models can explain seemingly detrimental shifts in policy on one dimension in favor of gains on another dimension. In conventional public choice models, a political actor will only give way to a loss in utility on one dimension if it entails an overall gain in utility in multidimensional space. A fuzzy public choice model, on the contrary, allows actor's to make compromises on one dimension for gains in another without necessarily experiencing a loss in utility on the former. In this way, scholars can predict whether an actor may be more likely to make a compromise for a gain, when the actor perceives no loss in utility on the compromised dimension, while the actor may be more resistant to such a trade-off if some real loss in utility occurs as a result.

In the next chapter we address issues related to fuzzy single-dimensional public choice models. The discussion will help us to present and test our fuzzy single-dimensional public choice models of government formation in Chap. 4.

References

Arrow, K.: Social Choice and Individual Values. Wiley, New York (1951)

Austen-Smith, D., Banks, J.: Positive Political Theory I. The University of Michigan Press, Ann Arbor (1999)

Benoit, K., Laver, M., Mikhalov, S.: Treating words as data with error: Estimating uncertainty in the comparative manifesto measures. Am. J. Polit. Sci. **53**(2), 49–513 (2009)

Black, D.: The Theory of Committees and Elections. Cambridge University Press, Cambridge (1958)

Budge, I., Klingemann, H.D., Volkens, A., Bara, J., Tannenbaum, E. (eds.): Mapping Policy Preferences: Estimates for Parties, Elections, and Governments: 1945–1998. Oxford University Press, Oxford (2001)

Klingemann, H.D., Volkens, A., Bara, J., Budge, I., McDonald, M.: Mapping Policy Prefences II: Estimates for Parties, Electors, and Governments in Eastern Europe, European Union, and OECD 1990–2003. Oxford University Press, Oxford (2006)

Laver, M.: How should we estimate the policy positions of political actors. In: Estimating the Policy Position of Political Actors. Routledge, London (2001)

Laver, M., Budge, I. (eds.): Party Policy and Goverment Coalitions. MacMillan Press, London (1992)

Mikhaylov, S., Laver, M., Benoit, K.: Coder reliability and misclassification in the human coding of party manifestos. Polit. Anal. **20**(1), 78–91 (2012)

Poole, K.: Spatial Model of Parliamentary Voting. Cambridge University Press, Cambridge (2005)

Poole, K., Rosenthal, H.: Ideology and Congress. Transaction Publishers, New Brunswick, NJ (2007)

Potter, J.: Extracting fuzzy preferences from empirical data (2007), unpublished thesis

Chapter 3
Fuzzy Single-Dimensional Public Choice Models

Abstract We consider the problem of intransitivity in collective preference in the presence of which there is no maximal set upon which to base a prediction of an outcome founded on collective preference. We give special attention to the conditions identified by Black's Median Voter Theorem that guarantee against intransitivity and assure a maximal set. We then argue that the theorem holds when preferences are fuzzy.

3.1 Intransitivity in Collective Preference

Intransitivity in collective preferences presents a major issue for public choice models. Intransitivity results in an empty maximal set, that is a set of alternatives each of which is not strictly dominated by some other alternative. Formally, $M(R, X) = \{x \in X \mid \forall y \in X, xRy\}$, where X is a set of alternatives and R is a binary relation on X. The rules used to determine the binary relation are otherwise known as aggregation rules. Thus, if the maximal set is empty, every alternative can be defeated in a head-to-head contest by at least one other alternative under any aggregation rules. The problem, which is referred to as majority cycling when majority rule is the aggregation rule used to determine the winner of such head-to-head contests, renders public choice models relying on social preference ordering unable to make a prediction (McKelvey 1976).

Black's Median Voter Theorem (Black 1958) identifies an important class of exceptions. The theorem demonstrates that when social preferences are restricted to those that are single-peaked in single-dimensional models, a non-empty maximal set is guaranteed under any set of aggregation rules. In other words, if the alternatives can be aligned along a single continuum in a manner such that the preferences of each player descend monotonically from their respective ideal points, then the maximal set comprises the subset of alternative at the median along that continuum.

P. C. Casey et al., *Fuzzy Social Choice Models*, Studies in Fuzziness and Soft Computing 318, DOI: 10.1007/978-3-319-08248-6_3, © Springer International Publishing Switzerland 2014

Not surprisingly, Black's Median Voter Theorem has enjoyed wide use in public choice models. It has seen particular use in models predicting the outcome of presidential vetoes and the likelihood that a legislative committee will report a bill to the floor of the parent assembly. (See for example Downs 1975; Kiewit and McCubbins 1988).

In this chapter, we demonstrate that Black's Median Voter Theorem holds for fuzzy single-dimensional public choice models. Much of what we discuss in this chapter has appeared in Gibilisco et al. (2012) and Mordeson et al. (2010). We repeat the argument in this earlier work for the sake of clarity and the completeness of our argument. As we will show, Black's Median Voter Theorem adheres in the fuzzy public choice model despite significant differences between conventional models and their fuzzy counterparts.

3.2 Fuzzy Aggregation Preference Rules

In this section, we discuss properties of rules for determining the social ordering of alternatives, or social preference, based on the individual preferences of those in a collective (political) setting. The public choice literature refers to these as aggregation rules.

Let $N = \{1, ..., n\}$ be a set of individuals, $n \geq 2$, and X a finite set of alternatives such that $|X| \geq 3$. A weak order on X is a binary relation R on X such that R is reflexive, complete, and transitive. Let $\mathcal{R}$ denote the set of all weak orders on X and $\mathcal{B}$ the set of all reflexive and complete binary relations on X. Let $\mathcal{R}^n = \{(R_1, ..., R_n) \mid R_i \in \mathcal{R}, i = 1, ..., n\}$. Let $\mathcal{FR}^*(X)$ denote the set of all binary fuzzy relations on X that are not empty and $\mathcal{FP}^*(X)$ the set of all fuzzy subsets of X that are not empty.

A **weak order** on X is a fuzzy binary relation ρ on X that is reflexive, complete, and (max–min) transitive. Assume that each individual i has a weak order ρ_i on X. Let $\mathcal{FR}$ denote the set of all weak orders on X.

Definition 3.1 (*fuzzy preference profile*) A **fuzzy preference profile** on X is a n-tuple of weak orders $\rho = (\rho_1, ..., \rho_n)$ describing the preferences of all individuals. Let $\mathcal{FR}^n$ denote the set of all fuzzy preference profiles. For all $\rho \in \mathcal{FR}^n$ and $\forall x, y \in X$, let

$$P(x, y; \rho) = \{i \in N \mid \pi_i(x, y) > 0\}$$

and

$$R(x, y; \rho) = \{i \in N \mid \rho_i(x, y) > 0\} \ .$$

Let $\mathcal{FB}$ denote the set of all reflexive and complete fuzzy binary relations on X.

Definition 3.2 (*Fuzzy Preference Aggregation Rule*) A function

$$\tilde{f} : \mathcal{FR}^n \to \mathcal{FB}$$

is called a **fuzzy preference aggregation rule**.

If $\widetilde{f}$ is a fuzzy preference aggregation rule and $\rho \in \mathscr{FR}^n$, we sometimes suppress $\widetilde{f}$ and write ρ for $\widetilde{f}(\rho)$ and $\pi(x, y) > 0$ for $\widetilde{f}(\rho)(x, y) > 0, \widetilde{f}(\rho)(y, x) = 0$.

Definition 3.3 (*Coalitions*) Let $\widetilde{f}$ be a fuzzy preference aggregation rule. Let λ be a fuzzy subset of N. Then λ is called a **coalition** if $|\mathrm{Supp}(\lambda)| \geq 2$.

Definition 3.4 Let $\widetilde{f}$ be a fuzzy preference aggregation rule. Let $(x, y) \in X \times X$. Let λ be a fuzzy subset of N. Then

(1) λ is called **semidecisive** for x against y, written $x\widetilde{D}_\lambda y$, if $\forall \rho \in \mathscr{FR}^n$,

$$[\pi_i(x, y) > 0 \ \forall i \in \mathrm{Supp}(\lambda)$$
$$\text{and } \pi_j(y, x) > 0 \ \forall j \notin \mathrm{Supp}(\lambda)] \text{ implies } \pi(x, y) > 0;$$

(2) λ is called **decisive** for x against y, written $xD_\lambda y$, if $\forall \rho \in \mathscr{FR}^n$, $[\pi_i(x, y) > 0 \forall i \in \mathrm{Supp}(\lambda)]$ implies $\pi(x, y) > 0$.

Definition 3.5 Let $\widetilde{f}$ be a fuzzy preference aggregation rule. Let λ be a fuzzy subset of N. Then λ is called **semidecisive** (**decisive**) if $\forall (x, y) \in X \times X$, λ is semidecisive (decisive) for x against y.

Definition 3.6 Let $\widetilde{f}$ be a fuzzy preference aggregation rule. Let $\mathscr{L}(\widetilde{f}) = \{\lambda \in \mathscr{FP}(N) \mid \lambda \text{ is decisive for } \widetilde{f}\}$.

Definition 3.7 Let $\widetilde{f}$ be a fuzzy preference aggregation rule. Then

(1) $\widetilde{f}$ is said to be **nondictatorial** if there does not exist $i \in N$ (the dictator) such that $\forall \rho \in \mathscr{FR}^n$, $\forall x, y \in X$, $\pi_i(x, y) > 0$ implies $\pi(x, y) > 0$;
(2) $\widetilde{f}$ is said to be **weakly Paretian** if $\forall \rho \in \mathscr{FR}^n$, $\forall x, y \in X$, $[\pi_i(x, y) > 0 \ \forall i \in N$ implies $\pi(x, y) > 0]$.

Proposition 3.8 *Let $\widetilde{f}$ be a fuzzy preference aggregation rule. Suppose $\widetilde{f}$ is dictatorial with dictator i. Let λ be a fuzzy subset of N. Suppose λ is a coalition. Then λ is decisive if and only if $i \in \mathrm{Supp}(\lambda)$.*

Proof Suppose λ is decisive. Suppose $i \notin \mathrm{Supp}(\lambda)$. Let $x, y \in X$ be such that $x \neq y$. Then $\exists \rho \in \mathscr{FR}^n$ such that $\pi_j(x, y) > 0 \forall j \in \mathrm{Supp}(\lambda)$ and $\pi_i(y, x) > 0$. Thus $\pi(x, y) > 0$ and $\pi(y, x) > 0$, a contradiction. Hence $i \in \mathrm{Supp}(\lambda)$. The converse is immediate.

Thus, if an aggregation rule results in the collective social preference being that of one individual i, then i is a dictator, and a coalition λ is decisive if and only if individual i is at least partially a member of λ.

Definition 3.9 Let $\widetilde{f}$ be a fuzzy preference aggregation rule. Define $f : \mathscr{R}^n \to \mathscr{B}$ by $\forall \rho = (R_1, ..., R_n) \in \mathscr{R}^n$,

$$f(\rho) = \left\{(x, y) \in X \times X \mid \widetilde{f}(1_{R_1}, ..., 1_{R_n})(x, y) > 0\right\} .$$

Then f is called the **preference aggregation rule** associated with $\widetilde{f}$.

Note that $\forall (x, y) \in X \times X$,

$$(x, y) \in f(R_1, ..., R_n) \Leftrightarrow (x, y) \in \text{Supp}(\widetilde{f}(1_{R_1}, ..., 1_{R_n})).$$

Hence $\forall (x, y) \in X \times X$,

$$(x, y) \in \text{Supp}(\widetilde{f}(1_{\text{Supp}(\rho_1)}, ..., 1_{\text{Supp}(\rho_n)})) \Leftrightarrow (x, y) \in f(\text{Supp}(\rho_1), ..., \text{Supp}(\rho_n)).$$

Let $\widetilde{g} = \widetilde{f}(1_{\text{Supp}(\rho_1)}, ..., 1_{\text{Supp}(\rho_n)})$, thus

$$\begin{aligned}
\{i \in N \mid \widetilde{g}(y, x) > 0, \widetilde{g}(x, y) = 0\} &= \{i \in N \mid (x, y) \in \text{Supp}(\widetilde{g}), (y, x) \notin \text{Supp}(\widetilde{g})\} \\
&= \{i \in N \mid (x, y) \in f(\text{Supp}(\rho_1), ..., \text{Supp}(\rho_n)), \\
&\qquad (y, x) \notin f(\text{Supp}(\rho_1), ..., \text{Supp}(\rho_n))\} \,.
\end{aligned}$$

Since $\widetilde{f}(1_{R_1}, ..., 1_{R_n}) \in \mathscr{FB}$, $(x, x) \in \text{Supp}(\widetilde{f}(1_{R_1}, ..., 1_{R_n}))$ and so $(x, x) \in f(\rho)$ for all $x \in X$. Also, $\forall x, y \in X$, either $(x, y) \in f(\rho)$ or $(y, x) \in f(\rho)$ since either $(x, y) \in \text{Supp}(\widetilde{f}(1_{R_1}, ..., 1_{R_n}))$ or $(y, x) \in \text{Supp}(\widetilde{f}(1_{R_1}, ..., 1_{R_n}))$ because $\widetilde{f}(1_{R_1}, ..., 1_{R_n}) \in \mathscr{FB}$. That is, $f(\rho) \in \mathscr{B}$.

Let $\rho \in \mathscr{FR}^n$. Let $\widetilde{f}(\rho)$ be denoted by ρ. Let $\rho \in R^n$ and $f(\rho)$ be denoted by R. Let $\widetilde{\rho} = (1_{R_1}, ..., 1_{R_n})$, where $\rho = (R_1, ..., R_n)$. Then $P(x, y; \widetilde{\rho}) = P(x, y, \rho)$ since $1_{R_i}(x, y) > 0$ if and only if $(x, y) \in R_i, i = 1, ..., n$.

Proposition 3.10 *Let $\widetilde{f}$ be a fuzzy preference aggregation rule. Let λ be a fuzzy subset of N. If λ is decisive with respect to $\widetilde{f}$, then $\text{Supp}(\lambda)$ is decisive with respect to f, where f is the preference aggregation rule associated with $\widetilde{f}$.*

Proof Let $\rho = (R_1, ..., R_n) \in \mathscr{R}^n$. Let $(x, y) \in X \times X$. Suppose $(x, y) \in P_i \, \forall i \in \text{Supp}(\lambda)$. Then $1_{P_i}(x, y) > 0 \, \forall i \in \text{Supp}(\lambda)$. Since λ is decisive with respect $\widetilde{f}$, $\pi(x, y) > 0$. That is, $\widetilde{f}(1_{R_1}, ..., 1_{R_n})(x, y) > 0$ and $\widetilde{f}(1_{R_1}, ..., 1_{R_n})(y, x) = 0$. Hence $(x, y) \in \text{Supp}(\widetilde{f}(1_{R_1}, ..., 1_{R_n}))$ and $(y, x) \notin \text{Supp}(\widetilde{f}(1_{R_1}, ..., 1_{R_n}))$. Thus by comments following Definition 3.9, $(x, y) \in f(R_1, ..., R_n)$ and $(y, x) \notin f(R_1, ..., R_n)$. Hence xPy.

Definition 3.11 Let $\widetilde{f}$ be a fuzzy preference aggregation rule. Then

(1) $\widetilde{f}$ is called a **simple majority rule** if $\forall \rho \in \mathscr{FR}^n$, $\forall x, y \in X$, $\pi(x, y) > 0$ if and only if $|P(x, y; \rho)| > n/2$;
(2) $\widetilde{f}$ is called a **Pareto extension rule** if $\forall \rho \in \mathscr{FR}^n$, $\forall x, y \in X$, $\pi(x, y) > 0$ if and only if $R(x, y; \rho) = N$ and $P(x, y, \rho) \neq \emptyset$.

Proposition 3.12 *Let $\widetilde{f}$ be a fuzzy preference aggregation rule. Let λ be a fuzzy subset of N. Suppose $\widetilde{f}$ is a simple majority rule. If λ is decisive with respect to $\widetilde{f}$, then $|Supp(\lambda)| > \frac{n}{2}$.*

Proof Since $\widetilde{f}$ is a simple majority rule, it follows by comments preceding Proposition 3.10 that f is a simple majority rule. Since λ is decisive with respect to $\widetilde{f}$,

Supp(λ) is decisive with respect to f, where f is the aggregation rule associated with $\widetilde{f}$ by Proposition 3.10. Hence $|\text{Supp}(\lambda)| > \frac{n}{2}$ by Austen-Smith and Banks (Example 2.5, p. 35).

Proposition 3.13 *Let $\widetilde{f}$ be a fuzzy preference aggregation rule. Let λ be a fuzzy subset of N. Suppose f is a Pareto extension rule. If λ is decisive with respect to $\widetilde{f}$, then $Supp(\lambda) = N$.*

Proof Since $\widetilde{f}$ is a Pareto extension rule, it follows by comments preceding proposition 3.11 that f is a Pareto extension rule. Since λ is decisive with respect to $\widetilde{f}$, Supp(λ) is decisive with respect to f, where f is the aggregation rule associated with $\widetilde{f}$ by Proposition 3.10. The desired result follows by Austen-Smith and Banks (1999, Example 2.5, p. 35).

We now look more carefully at sets of decisive coalitions.

Definition 3.14 Let $\mathscr{L} \subseteq \mathscr{F}\mathscr{P}(N)$ be such that $\forall \lambda \in \mathscr{L}$, λ is a coalition. Then

(1) $\mathscr{L}$ is said to be **monotonic** if $\forall \lambda \in \mathscr{L}$, $\forall \lambda' \in \mathscr{F}\mathscr{P}(N)$, $N \supseteq \text{Supp}(\lambda') \supset \text{Supp}(\lambda)$ implies $\lambda' \in \mathscr{L}$;
(2) $\mathscr{L}$ is said to be **proper** if $\forall \lambda \in \mathscr{L}$, $\forall \lambda' \in \mathscr{F}\mathscr{P}(N)$, $\text{Supp}(\lambda') \cap \text{Supp}(\lambda) = \emptyset$ implies $\lambda' \notin \mathscr{L}$.

Proposition 3.15 *Let $\widetilde{f}$ be a fuzzy preference aggregation rule. Then*

(1) *$L(\widetilde{f})$ is monotonic;*
(2) *$L(\widetilde{f})$ is proper.*

Proof (1) Suppose $\lambda, \lambda' \in \mathscr{F}\mathscr{P}(N)$ are such that $\text{Supp}(\lambda) \subset \text{Supp}(\lambda')$. Suppose λ' is not decisive for $\widetilde{f}$. Then there exists $\rho \in \mathscr{F}\mathscr{R}^n$ and $x, y \in X$ such that $\pi_i(x, y) > 0$ for all $i \in \text{Supp}(\lambda')$, but not $\pi(x, y) > 0$. Thus $\pi_i(x, y) > 0$ for all $i \in \text{Supp}(\lambda)$, but not $\pi(x, y) > 0$. Hence λ is not decisive for $\widetilde{f}$.

(2) Let $\lambda \in \mathscr{L}(\widetilde{f})$. Let $x, y \in X$ and $\rho \in \mathscr{F}\mathscr{R}^n$ be such that $\pi_i(x, y) > 0$ for all $i \in \text{Supp}(\lambda)$ and $\pi_i(y, x) > 0$ for all $i \in N \backslash \text{Supp}(\lambda)$. Then $\pi(x, y) > 0$. Let $\lambda' \in FP(N)$. Suppose that $\text{Supp}(\lambda) \cap \text{Supp}(\lambda') = \emptyset$. If $\lambda' \in \mathscr{L}(\widetilde{f})$, then $\pi(y, x) > 0$, a contradiction. Hence λ' is not decisive for $\widetilde{f}$ for all $\lambda' \in \mathscr{F}\mathscr{P}(N)$ such that $\text{Supp}(\lambda') \cap \text{Supp}(\lambda) = \emptyset$. Thus $\mathscr{L}(\widetilde{f})$ is proper.

Definition 3.16 *(Collegial)* Let $\widetilde{f}$ be a fuzzy aggregation rule. If $\forall \mathscr{L}^*(\widetilde{f}) \subseteq \mathscr{L}(\widetilde{f})$ such that $0 < |\mathscr{L}^*(\widetilde{f})| < \infty$, $\bigcap_{\lambda \in \mathscr{L}^*(\widetilde{f})} \lambda \neq \theta$, then $\widetilde{f}$ is called **collegial**.

The fuzzy collegial rule provides a powerful concept for predicting outcomes in public choice models. Members of the fuzzy collegium are members of every potential winning coalition to some degree. Thus, no final decision can be reached without their concurrence. This permits us to focus on those alternatives favored by members of the fuzzy collegium As we will demonstrate, the political actors whose most preferred alternatives lie at the median comprise the fuzzy collegium in single peaked, single-dimensional public choice models. Thus Black's Median Voter Theorem holds with fuzzy preferences.

Definition 3.17 *(Acyclicity)* Let $\rho \in \mathscr{FR}$. Then

(1) ρ is called **acyclic** if $\forall x_1, ..., x_n \in X$, we have that

$$\pi(x_1, x_2) \wedge ... \wedge \pi(x_{n-1}, x_n) \leq \rho(x_1, x_n) \ ;$$

(2) ρ is called **partially acyclic** if $\forall x_1, ..., x_n \in X$,

$$\pi(x_1, x_2) > 0, ..., \pi(x_{n-1}, x_n) > 0 \text{ implies } \rho(x_1, x_n) > 0.$$

Definition 3.18 *(Partial Acyclicity)* Let $\widetilde{f}$ be a fuzzy preference aggregation rule. Then $\widetilde{f}$ is called **partially acyclic** if $\widetilde{f}(\rho)$ is partially acyclic $\forall \rho \in \mathscr{FR}^n$.

Proposition 3.19 *Let $\widetilde{f}$ be a fuzzy aggregation rule and f the aggregation rule associated with $\widetilde{f}$.*

(1) *If $\widetilde{f}$ is partially acyclic, then f is acyclic.*
(2) *If $\widetilde{f}$ is weakly Paretian, then f is weakly Paretian.*

Proof (1) Since $\widetilde{f}$ is partially acyclic, it follows that $\forall x_1, ..., x_m \in X$, $1_P(x_1, x_2) > 0, ..., 1_P(x_{m-1}, x_m) > 0$ implies $1_R(x_1, x_m) > 0$, where P and R correspond to f. Thus $x_1 P x_2, ..., x_{m-1} P x_m$ implies $x_1 R x_m$ since $1_P(x_{i-1}, x_i) > 0 \Leftrightarrow (x_{i-1}, x_i) \in \text{Supp}(1_P) = P, i = 2, ..., n$.

   (2) Since $\widetilde{f}$ is weakly Paretian, $\forall \rho \in \mathscr{R}^n$, it follows that $\forall x, y \in X, (\forall i \in N, 1_{P_i}(x, y) > 0$ implies $1_P(x, y) > 0)$. Thus $\forall i \in N, x P_i y$ implies $x P y$.

   The following result states in words that any fuzzy preference aggregation rule that is partially acyclic and weakly Paretian results in a fuzzy collegium.

Proposition 3.20 *Let $\widetilde{f}$ be a fuzzy aggregation rule. If $\widetilde{f}$ is partially acyclic and weakly Paretian, then f is collegial.*

Proof By Proposition 3.19, f is acyclic and weakly Paretian. Thus

$$\bigcap_{\lambda \in \mathscr{L}(\widetilde{f})} \text{Supp}(\lambda) \neq \emptyset$$

by Proposition 3.10 and Austen-Smith and Banks (1999, Theorem 2.4, p. 43). Hence

$$\bigcap_{\lambda \in \mathscr{L}^*(\widetilde{f})} \lambda \neq \theta \ \forall \mathscr{L}^*(\widetilde{f}) \subseteq \mathscr{L}(\widetilde{f})$$

such that $0 < \left| \mathscr{L}^*(\widetilde{f}) \right| < \infty$.

3.3 Fuzzy Voting Rules

We now turn to a consideration of fuzzy preference aggregation rules. In the conventional approach, all preference aggregation rules are voting rules Austen-Smith and Banks. Moreover, simple rules constitute an important subset of voting rules. These rules include majority rules, super majority rules, and the unanimity rule.

We proceed by characterizing fuzzy voting rules and fuzzy simple rules. We conclude with our major result (Theorem 3.43): fuzzy simple rules produce a maximal set located at the median, which is generalizable to the core in n-dimensional space, $n > 1$. Thus, when fuzzy simple rules are used to aggregate the collective preferences of actors, fuzzy single-dimensional public choice models predict an outcome at the median.

Let $R \subseteq X \times X$. We let Symm (R) denote the symmetric closure of R, i. e., the smallest subset of $X \times X$ that contains R and is symmetric.

Definition 3.21 Let $\widetilde{f}$ be a fuzzy aggregation rule. Let $\mathscr{L} \subseteq \mathscr{FP}(N)$. Let $\rho \in \mathscr{FR}^n$ and set $\widetilde{\mathscr{P}}(\rho) = \{(x, y) \in X \times X \mid \exists \lambda \in \mathscr{L}, \forall i \in \text{Supp}(\lambda), \ \pi_i(x, y) > 0\}$. Define $\widetilde{f}_{\mathscr{L}} : \mathscr{FR}^n \to \mathscr{FB}$ by $\forall \rho \in \mathscr{FR}^n, \forall x, y \in X$,

$$
\widetilde{f}_{\mathscr{L}}(\rho)(x, y) = \begin{cases} 1 & \text{if } (x, y) \notin \text{Symm}(\widetilde{\mathscr{P}}(\rho)), \\ \bigvee \{\pi_i(x, y) \mid \exists \lambda \in \mathscr{L}, \forall i \in \text{Supp}(\lambda), \pi_i(x, y) > 0\} \\ & \text{if } (x, y) \in \widetilde{\mathscr{P}}(\rho), \\ 0 & \text{if } (x, y) \in \text{Symm}(\widetilde{\mathscr{P}}(\rho)) \backslash \widetilde{\mathscr{P}}(\rho). \end{cases}
$$

The following example illustrates the connections of $\widetilde{f}_{\mathscr{L}}$ to the conventional approach (Austen-Smith and Banks 1999, p. 59).

Example 3.22 Let $X = \{x, y, z\}, N = \{1, 2, 3\}$, and $\rho = (\rho_1, \rho_2, \rho_3)$, where $\rho_i(x, x) = \rho_i(y, y) = \rho_i(z, z) = 1, i = 1, 2, 3$ and

$$
\rho_1(x, y) = \frac{3}{4}, \qquad \rho_1(x, z) = \frac{1}{4}, \qquad \rho_1(y, z) = \frac{1}{2},
$$
$$
\rho_2(x, y) = \frac{1}{2}, \qquad \rho_2(x, z) = \frac{1}{3}, \qquad \rho_2(z, y) = \frac{2}{3},
$$
$$
\rho_3(y, x) = \frac{3}{4}, \qquad \rho_3(z, x) = \frac{1}{2}, \qquad \rho_3(y, z) = \frac{1}{4}.
$$

Each ρ_i is defined to be 0 otherwise, $i = 1, 2, 3$. Let $\mathscr{L} = \{\lambda, \lambda'\}$, where $\lambda(1) = \frac{1}{2}$, $\lambda(2) = \frac{1}{4}$, $\lambda(3) = 0$ and $\lambda'(1) = 0, \lambda'(2) = \frac{3}{4}, \lambda'(3) = \frac{1}{2}$. Now $\text{Supp}(\lambda) = \{1, 2\}$ and $\text{Supp}(\lambda') = \{2, 3\}$. Hence $\widetilde{\mathscr{P}}(\rho) = \{(x, y), (x, z)\}$ and $\text{Symm}(\widetilde{\mathscr{P}}(\rho)) = \{(x, y), (y, x), (x, z), (z, x)\}$. Thus it follows that

$$\tilde{f}_{\mathscr{L}}(\rho)(x, y) = \frac{3}{4} \vee \frac{1}{2}$$
$$= \frac{3}{4},$$

$$\tilde{f}_{\mathscr{L}}(\rho)(x, z) = \frac{1}{4} \vee \frac{1}{3}$$
$$= \frac{1}{3}.$$

It also follows that $\tilde{f}_{\mathscr{L}}(\rho)(y, z) = \tilde{f}_{\mathscr{L}}(\rho)(z, y) = 1, \tilde{f}_{\mathscr{L}}(\rho)(y, x) = \tilde{f}_{\mathscr{L}}(\rho)(z, x) = 0$, and $\tilde{f}_{\mathscr{L}}(\rho)(x, x) = \tilde{f}_{\mathscr{L}}(\rho)(y, y) = \tilde{f}_{\mathscr{L}}(\rho)(z, z) = 1$.

Definition 3.23 (*Partial Fuzzy Simple Rule*) Let $\tilde{f}$ be a fuzzy aggregation rule. Then $\tilde{f}$ is called a **partial fuzzy simple rule** if $\forall \rho \in \mathscr{FR}^n, \forall x, y \in X, \tilde{f}(\rho)(x, y) > 0$ if and only if $\tilde{f}_{\mathscr{L}(\tilde{f})}(\rho)(x, y) > 0$.

Proposition 3.24 *Let $\tilde{f}$ be a fuzzy preference aggregation rule and let f be the aggregation rule associated with $\tilde{f}$. If $\tilde{f}$ is a partial fuzzy simple rule, then f is a simple rule.*

Proof Let $\rho \in \mathscr{FR}^n$ and let $x, y \in X$. Recall that $\widetilde{\mathscr{P}}(\rho) = \{(x, y) \in X \times X \mid \exists \lambda \in \mathscr{L}(\tilde{f}), \forall i \in \mathrm{Supp}(\lambda), \pi_i(x, y) > 0\}$ and $\mathscr{L}(\tilde{f}) = \{\lambda \in \mathscr{FP}(N) \mid \lambda$ is decisive for $\tilde{f}\} = \{\lambda \in \mathscr{FP}(N) \mid \forall \rho \in \mathscr{FR}^n, \pi_i(x, y) > 0 \; \forall i \in \mathrm{Supp}(\lambda) \Rightarrow \pi(x, y) > 0\}$. Let $\rho = (1_{R_1}, ..., 1_{R_n})$, where $R_i \in \mathscr{R}, i = 1, ..., n$. Then,

$$\tilde{f}_{\mathscr{L}(\tilde{f})}(\rho)(x, y) = \begin{cases} 1 \text{ if} (x, y) \notin \mathrm{Symm}(\widetilde{\mathscr{P}}(\rho)), \\ \bigvee\{1_{P_i}(x, y) \mid \exists \lambda \in \mathscr{L}(\tilde{f}), \forall i \in \mathrm{Supp}(\lambda), 1_{P_i}(x, y) = 1\} \\ \quad \text{if} (x, y) \in \widetilde{\mathscr{P}}(\rho), \\ 0 \text{ if} (x, y) \in \mathrm{Symm}(\widetilde{\mathscr{P}}(\rho)) \backslash \widetilde{\mathscr{P}}(\rho). \end{cases}$$

Thus for $\rho = (1_{R_1}, ..., 1_{R_n})$,

$$\tilde{f}_{\mathscr{L}(\tilde{f})}(\rho)(x, y) = \begin{cases} 1 \text{ if} (x, y) \in \widetilde{\mathscr{P}}(\rho) \cup (X \times X \backslash \mathrm{Symm}(\widetilde{\mathscr{P}}(\rho))), \\ 0 \text{ if} (x, y) \in \mathrm{Symm}(\widetilde{\mathscr{P}}(\rho)) \backslash \widetilde{\mathscr{P}}(\rho). \end{cases}$$

Now

$$f(R_1, ... R_n) = \{(x, y) \mid \tilde{f}(1_{R_1}, ..., 1_{R_n})(x, y) > 0\}$$
$$= \mathrm{Supp}(\tilde{f}(1_{R_1}, ..., 1_{R_n}))$$
$$= \mathrm{Supp}(\tilde{f}_{\mathscr{L}(\tilde{f})}(1_{R_1}, ..., 1_{R_n}))$$
$$= \widetilde{\mathscr{P}}(\rho) \cup (X \times X \backslash \mathrm{Symm}(\widetilde{\mathscr{P}}(\rho))).$$

By definition,

$$f_{\mathcal{L}(f)}(R_1, ..., R_n) = \mathcal{P}(\rho) \cup (X \times X \backslash \mathrm{Symm}(\mathcal{P}(\rho))),$$

where $\mathcal{P}(\rho) = \{(x, y) \mid \exists L \in \mathcal{L}(f), \forall i \in L, xP_iy\}$ and $\rho = (R_1, ..., R_n)$. Now for $\rho = (1_{R_1}, ..., 1_{R_n})$, $\exists \lambda \in \mathcal{L}(\tilde{f})$, $\forall i \in \mathrm{Supp}(\lambda)$, xP_iy (P_i corresponds to 1_{R_i}) $\Leftrightarrow$ $\exists L \in \mathcal{L}(f)$, $\forall i \in L$, xP_iy (P_i corresponds to R_i), i. e., given λ, let $L = \mathrm{Supp}(\lambda)$ and given L, let $\lambda = 1_L$. Thus $\tilde{\mathcal{P}}((1_{R_1}, ..., 1_{R_n})) = \mathcal{P}((R_1, ..., R_n))$. Hence $\tilde{\mathcal{P}}(1_{R_1}, ..., 1_{R_n}) = \mathcal{P}(R_1, ..., R_n)$. Thus $f = f_{\mathcal{L}(f)}$.

Definition 3.25 Let $\tilde{f}$ be a fuzzy aggregation rule. Then

(1) $\tilde{f}$ is called **decisive** if $\forall \rho, \rho' \in \mathcal{FR}^n$ and $\forall x, y \in X$,

$$\left[P(x, y; \rho) = P(x, y; \rho') \text{ and } \pi(x, y) > 0 \right]$$

imply $\pi'(x, y) > 0$;

(2) $\tilde{f}$ is called **monotonic** if $\forall \rho, \rho' \in \mathcal{FR}^n$ and $\forall x, y \in X$,

$$\left[P(x, y; \rho) \subseteq P(x, y; \rho'), R(x, y; \rho) \subseteq R(x, y; \rho') \text{ and } \pi(x, y) > 0 \right]$$

imply $\pi'(x, y) > 0$.

Definition 3.26 (*Neutrality*) Let $\tilde{f}$ be a fuzzy aggregation rule. Then $\tilde{f}$ is called **neutral** if $\forall \rho, \rho' \in \mathcal{FR}^n, \forall x, y, z, w \in X, P(x, y; \rho) = P(z, w; \rho')$ and $P(y, x; \rho) = P(w, z; \rho')$ imply $\tilde{f}(\rho)(x, y) > 0$ if and only if $\tilde{f}(\rho')(z, w) > 0$.

Theorem 3.27 *Let $\tilde{f}$ be a fuzzy aggregation rule. Then $\tilde{f}$ is a partial fuzzy simple rule if and only if $\tilde{f}$ is decisive, neutral, and monotonic.*

Proof Let $x, y \in X$. Suppose $\pi_{\mathcal{L}(\tilde{f})}(x, y) > 0$. Then $\exists \lambda \in \mathcal{L}(\tilde{f})$ such that $\forall i \in \mathrm{Supp}(\lambda), \pi_i(x, y) > 0$ and so $\pi(x, y) > 0$. Suppose $\tilde{f}$ is decisive, neutral, and monotonic. Let $\rho \in \mathcal{FR}^n$. Suppose $\pi(x, y) > 0$. In order to show $\tilde{f}$ is a partial fuzzy simple rule, it suffices to show $P(x, y; \rho) \in \mathcal{L}(\tilde{f})$ for then $\pi_{\mathcal{L}(\tilde{f})}(x, y) > 0$. Let $a, b \in X$. Let ρ^* be any fuzzy preference profile such that $\forall i \in P(x, y; \rho), \pi_i^*(a, b) > 0$. Let $L^+ = P(a, b; \rho^*) \backslash P(x, y; \rho)$. Let ρ^1 be a fuzzy preference profile such that $P(a, b; \rho^1) = P(x, y; \rho)$ and $P(b, a; \rho^1) = P(y, x; \rho)$. Since $\tilde{f}$ is neutral, $\pi^1(a, b) > 0$. Let $\rho^2 \in \mathcal{FR}^n$ be defined by $\rho_i^2|_{\{a,b\}} = \rho_i^*|_{\{a,b\}}$ if and only if $i \in L^+ \cup P(x, y; \rho)$ and $\rho_j^2|_{\{a,b\}} = \rho_j^1|_{\{a,b\}}$ otherwise. Thus individuals that ρ^2 and ρ^1 differ from 0 on a, b must come from L^+. Thus $\pi^2(a, b) > 0$ since $\tilde{f}$ is monotonic. Since $P(a, b; \rho^*) = P(a, b; \rho^2)$ and $\tilde{f}$ is decisive, $\pi^*(a, b) > 0$. Thus since ρ and a, b are arbitrary except for $\pi_i^*(a, b) > 0$ if $i \in P(x, y; \rho)$ and $\pi^*(a, b) > 0$, it follows that $P(x, y; \rho) \in \mathcal{L}(\tilde{f})$.

Conversely, suppose that $\tilde{f}$ is a partial fuzzy simple rule. Then $\tilde{f}$ is neutral since $\mathcal{L}(\tilde{f})$ is defined without regard to alternatives. Monotonicity follows from Proposition 3.15 and the definition of $\tilde{f}_{\mathcal{L}(\tilde{f})}$. That $\tilde{f}$ is decisive follows directly from the definitions of $\mathcal{L}(\tilde{f})$ and $\tilde{f}_{\mathcal{L}(\tilde{f})}$.

Definition 3.28 (*Decisive Structure*) Let $\widetilde{f}$ be a fuzzy aggregation rule. The **decisive structure** of $\widetilde{f}$, denoted $\mathscr{D}(\widetilde{f})$, is defined to be the set $\mathscr{D}(\widetilde{f}) = \{(\sigma, \omega) \in \mathscr{F}\mathscr{P}(N) \times \mathscr{F}\mathscr{P}(N) \mid \text{Supp}(\sigma) \subseteq \text{Supp}(\omega) \text{ and } \forall x, y \in X, \forall \rho \in \mathscr{F}\mathscr{R}^n, \pi_i(x, y) > 0 \; \forall i \in \text{Supp}(\sigma) \text{ and } \rho_j(x, y) > 0 \forall j \in \text{Supp}(\omega) \text{ implies } \pi(x, y) > 0\}$.

An aggregation rule is a **plurality rule** if $\forall x, y \in X$, alternative x is chosen over alternative y when the number of individuals strictly preferring x to y exceeds the number strictly preferring y to x. The plurality rule is a voting rule; it is not a simple rule. Majority rule is both a simple rule and a voting rule.

Definition 3.29 (*Monotonic*) Let $\mathscr{D} \subseteq \mathscr{F}\mathscr{P}(N) \times \mathscr{F}\mathscr{P}(N)$. Then $\mathscr{D}$ is called **monotonic** if $(\sigma, \omega) \in \mathscr{D}$, $\text{Supp}(\sigma) \subseteq \text{Supp}(\sigma') \subseteq \text{Supp}(\omega')$ and $\text{Supp}(\sigma) \subseteq \text{Supp}(\omega) \subseteq \text{Supp}(\omega')$ imply $(\sigma', \omega') \in \mathscr{D}$.

Definition 3.30 Let $\widetilde{f}$ be a fuzzy aggregation rule. Let $\mathscr{D} \subseteq \mathscr{F}\mathscr{P}(N) \times \mathscr{F}\mathscr{P}(N)$ be such that $(\sigma, \omega) \in \mathscr{D}$ implies $\text{Supp}(\sigma) \subseteq \text{Supp}(\omega)$. Let $\rho \in \mathscr{F}\mathscr{R}^n$ and set

$$\mathscr{F}\mathscr{P}_{\mathscr{D}}(\rho) = \{(x, y) \in X \times X \mid \exists(\sigma, \omega) \in \mathscr{D}, \forall i \in \text{Supp}(\sigma), \; \pi_i(x, y) > 0$$
$$\text{and } \forall j \in \text{Supp}(\omega), \; \rho_j(x, y) > 0\} \;.$$

Define $\widetilde{f}_{\mathscr{D}} : \mathscr{F}\mathscr{R}^n \to \mathscr{F}\mathscr{B}$ by $\forall \rho \in \mathscr{F}\mathscr{R}^n, \forall x, y \in X$,

$$\widetilde{f}_{\mathscr{D}}(\rho)(x, y) = \begin{cases} 1 & \text{if } (x, y) \notin \text{Symm}(\mathscr{F}\mathscr{P}_{\mathscr{D}}(\rho)) \\ \bigvee \{\pi_i(x, y) \wedge \rho_j(x, y) \mid \exists(\sigma, \omega) \in \mathscr{D}, \\ \quad \pi_i(x, y) > 0, \; \forall i \in \text{Supp}(\sigma), & \text{if } (x, y) \in \mathscr{F}\mathscr{P}_{\mathscr{D}}(\rho), \\ \quad \rho_j(x, y) > 0 \; \forall j \in \text{Supp}(\rho_j)\} \\ 0 & \text{if } (x, y) \in \text{Symm}(\mathscr{F}\mathscr{P}_D(\rho)) \backslash \mathscr{F}\mathscr{P}_D(\rho). \end{cases}$$

Example 3.31 Let $X = \{x, y, z\}, N = \{1, 2, 3\}$, and $\rho = (\rho_1, \rho_2, \rho_3)$, where $\rho_i(x, x) = \rho_i(y, y) = \rho_i(z, z) = 1, i = 1, 2, 3$ and

$$\rho_1(x, y) = \frac{3}{4}, \qquad \rho_1(x, z) = \frac{1}{4}, \qquad \rho_1(y, z) = \frac{1}{2},$$
$$\rho_2(x, y) = \frac{1}{2}, \qquad \rho_2(x, z) = \frac{1}{3}, \qquad \rho_2(z, y) = \frac{2}{3},$$
$$\rho_3(x, y) = \frac{1}{8}, \qquad \rho_3(y, x) = \frac{3}{4}, \qquad \rho_3(z, x) = \frac{1}{2}, \qquad \rho_3(y, z) = \frac{1}{4}.$$

Each ρ_i is defined to be 0 otherwise, $i = 1, 2, 3$. Let $\mathscr{D} = \{(\sigma, \omega), (\sigma', \omega')\}$, where $\sigma(1) = \frac{1}{2}, \sigma(2) = \frac{1}{4}, \sigma(3) = 0, \omega(1) = \frac{1}{2}, \omega(2) = \frac{1}{4}, \omega(3) = \frac{1}{3}$, $\sigma'(1) = 0, \sigma'(2) = \frac{3}{4}, \sigma'(3) = \frac{1}{2}$, and $\omega' = \sigma'$. Now $\text{Supp}(\sigma) = \{1, 2\}$, Supp and $\text{Supp}(\sigma') = \{2, 3\} = \text{Supp}(\omega')$. Now $\pi_i(x, y) > 0 \; \forall i \in \text{Supp}(\sigma)$ and $\rho_i(x, y) > 0 \; \forall i \in \text{Supp}(\omega)$. Thus $(x, y) \in \mathscr{F}\mathscr{P}_{\mathscr{D}}(\rho)$. Now $\pi_i(x, z) > 0 \forall i \in \text{Supp}(\sigma)$, but $\rho_3(x, z) = 0$ and $3 \in \text{Supp}(\omega)$. Hence $(x, z) \notin \mathscr{F}\mathscr{P}_{\mathscr{D}}(\rho)$. Note also that $(y, z) \notin \mathscr{F}\mathscr{P}_{\mathscr{D}}(\rho)$ since it is not the case that $\pi_i(y, z) > 0 \forall i \in \text{Supp}(\sigma)$

or $\pi_i'(y, z) > 0 \forall i \in \text{Supp}(\sigma')$. Thus it follows that $\mathscr{F}\mathscr{P}_{\mathscr{D}}(\rho) = \{(x, y)\}$, $\text{Symm}(\mathscr{F}\mathscr{P}_{\mathscr{D}}(\rho)) = \{(x, y), (y, x)\}$ and

$$\tilde{f}(\rho)(x, y) = \bigvee \{\pi_i(x, y) \wedge \rho_j(x, y) \mid i = 1, 2; j = 1, 2, 3\}$$
$$= \bigvee \{\frac{3}{4} \wedge \frac{3}{4}, \frac{3}{4} \wedge \frac{1}{2}, \frac{3}{4} \wedge \frac{1}{8}, \frac{1}{2} \wedge \frac{1}{4}, \frac{1}{2} \wedge \frac{1}{2}, \frac{1}{2} \wedge \frac{1}{8}\}$$
$$= \frac{3}{4}.$$

It also follows that $\tilde{f}_{\mathscr{D}}(\rho)(u, v) = 1$ if $(y, x) \neq (u, v) \neq (x, y)$ and that $\tilde{f}(y, x) = 0$.

Definition 3.32 Let $\tilde{f}$ be a fuzzy aggregation rule. Then $\tilde{f}$ is called a **partial fuzzy voting rule** if $\forall \rho \in \mathscr{F}\mathscr{R}^n$ and $\forall x, y \in X, \tilde{f}(\rho)(x, y) > 0$ if and only if $\tilde{f}_{\mathscr{D}(\tilde{f})}(\rho)(x, y) > 0$.

Proposition 3.33 *If $\tilde{f}$ is a partial fuzzy voting rule, then f is a voting rule.*

Proof Let $\rho = (R_1, ..., R_n) \in R^n$ and let $\rho = (1_{R_1}, ..., 1_{R_n})$. Then

$$(x, y) \in \mathscr{F}\mathscr{P}_{\mathscr{D}(\tilde{f})}(\rho) \Leftrightarrow \exists (\sigma, \omega) \in \mathscr{D}(\tilde{f}), \forall i \in \text{Supp}(\sigma), 1_{P_i}(x, y) = 1,$$
$$\forall j \in \text{Supp}(\omega), 1_{R_j}(x, y) = 1$$
$$\Leftrightarrow \exists (\sigma, \omega) \in \mathscr{D}(\tilde{f}), \forall i \in \text{Supp}(1_{\text{Supp}(\sigma)}), 1_{P_i}(x, y) = 1,$$
$$\forall j \in \text{Supp}(1_{\text{Supp}(\omega)}), 1_{R_j}(x, y) = 1$$
$$\Leftrightarrow \exists S, W \in \mathscr{D}(f), \forall i \in S, (x, y) \in P_i, \forall j \in W, (x, y) \in R_j,$$

where $S = \text{Supp}(1_{\text{Supp}}(\sigma)), W = \text{Supp}(1_{\text{Supp}}(\omega))$ for the implication $\Rightarrow$ and $\sigma = 1_S$ and $\omega = 1_W$ for the implication $\Leftarrow$. Hence

$$(x, y) \in f(R_1, ..., R_n)$$
$$\Leftrightarrow (x, y) \in \text{Supp}(\tilde{f}(1_{R_1}, ..., 1_{R_n})$$
$$\Leftrightarrow (x, y) \in (X \times X \backslash \text{Symm}(\mathscr{F}\mathscr{P}_{\mathscr{D}(\tilde{f})}(\rho))) \cup (\mathscr{F}\mathscr{P}_{\mathscr{D}(\tilde{f})}(\rho))$$
$$\Leftrightarrow (x, y) \in (X \times X \backslash \text{Symm}(\mathscr{P}_{\mathscr{D}(f)}(R_1, ..., R_n))) \cup (\mathscr{P}_{\mathscr{D}(f)}(R_1, ..., R_n))$$
$$\Leftrightarrow (x, y) \in f_{\mathscr{D}(f)}(R_1, ..., R_n).$$

Thus $f = f_{\mathscr{D}(f)}$.

Theorem 3.34 *A fuzzy aggregation rule is a partial fuzzy voting rule if and only if it is neutral and monotonic.*

Proof Suppose $\tilde{f}$ is a partial fuzzy voting rule. We show that $\tilde{f}$ is monotonic. We first note that $\mathscr{D}(\tilde{f})$ is monotonic. Let $(\sigma, \omega) \in \mathscr{D}(\tilde{f})$ and $\sigma', \omega' \in \mathscr{F}\mathscr{P}(N)$ be such that $\text{Supp}(\sigma) \subseteq \text{Supp}(\sigma') \subseteq \text{Supp}(\omega')$ and $\text{Supp}(\sigma) \subseteq \text{Supp}(\omega) \subseteq \text{Supp}(\omega')$. Since $\text{Supp}(\sigma) \subseteq \text{Supp}(\sigma')$ and $\text{Supp}(\omega) \subseteq \text{Supp}(\omega')$, clearly $(\sigma', \omega') \in \mathscr{D}(\tilde{f})$. Let

$\rho, \rho' \in \mathscr{FR}^n$ and $x, y \in X$ be such that $P(x, y; \rho) \subseteq P(x, y; \rho'), R(x, y; \rho) \subseteq R(x, y; \rho')$, and $\pi(x, y) > 0$. By Definitions 3.30 and 3.32, we have that since $\pi(x, y) > 0, \exists (\sigma, \omega) \in \mathscr{D}(\widetilde{f})$ such that $\text{Supp}(\sigma) \subseteq \text{Supp}(\omega)$ and $\forall x, y \in X, \forall \rho \in \mathscr{FR}^n, \pi_i(x, y) > 0 \forall i \in \text{Supp}(\sigma)$ and $\rho_j(x, y) > 0 \forall j \in \text{Supp}(\omega)$. By hypothesis,

$$\{i \in N \mid \pi_i(x, y) > 0\} \subseteq \{i \in N \mid \pi_i'(x, y) > 0\},$$
$$\{j \in N \mid \rho_j(x, y) > 0\} \subseteq \{j \in N \mid \rho_j'(x, y) > 0\}.$$

Since $(\sigma', \omega') \in \mathscr{D}(\widetilde{f}), \pi'(x, y) > 0$. Thus $\widetilde{f}$ is monotonic. We now show $\widetilde{f}$ is neutral. Let $\rho, \rho' \in \mathscr{FR}^n$ and $x, y, z, w \in X$ be such that

$$\{i \in N \mid \pi_i(x, y) > 0\} = \{i \in N \mid \pi_i'(z, w) > 0\},$$
$$\{i \in N \mid \pi_i(y, x) > 0\} = \{i \in N \mid \pi_i'(w, z) > 0\}.$$

Then it follows easily that $\pi_i(x, y) > 0 \Leftrightarrow \pi_i'(z, w) > 0 \forall i \in N$. Suppose $\widetilde{f}(\rho)(x, y) > 0$. Then $(x, y) \in (X \times X \backslash \text{Symm}(\mathscr{FP}_{\mathscr{D}(\widetilde{f})}(\rho)) \cup \mathscr{FP}_{\mathscr{D}(\widetilde{f})}(\rho)$, say $(x, y) \in \mathscr{FP}_{\mathscr{D}(\widetilde{f})}(\rho)$. Then $\exists (\sigma, \omega) \in \mathscr{D}(\widetilde{f})$ such that $\forall i \in \text{Supp}(\sigma), \pi_i(x, y) > 0$, $\forall j \in \text{Supp}(\omega), \rho_i(x, y) > 0\}$. By hypothesis, it follows that $(z, w) \in \mathscr{FP}_{\mathscr{D}(\widetilde{f})}(\rho')$. Hence $\widetilde{f}(\rho')(z, w) > 0$ and in fact $\pi'(z, w) > 0$. Suppose $(x, y) \in X \times X \backslash \text{Symm}(\mathscr{FP}_{\mathscr{D}(\widetilde{f})}(\rho))$. By the argument just given, $(z, w) \notin \mathscr{FP}_{\mathscr{D}(\widetilde{f})}(\rho')$ else $(x, y) \in \mathscr{FP}_{\mathscr{D}(\widetilde{f})}(\rho)$. Thus $(z, w) \in X \times X \backslash \mathscr{FP}_{\mathscr{D}(\widetilde{f})}(\rho')$. Hence $\widetilde{f}(\rho')(z, w) > 0$. Thus $\widetilde{f}$ is neutral.

For the converse, we first show that $\pi_{\mathscr{D}(\widetilde{f})}(x, y) > 0$ implies $\pi(x, y) > 0$. Suppose

$$\widetilde{f}_{\mathscr{D}(\widetilde{f})}(\rho)(x, y) > 0$$

for $\rho \in \mathscr{FR}^n$ and $x, y \in X$. Then

$$(x, y) \in \left(X \times X \backslash \text{Symm}(\mathscr{FP}_{\mathscr{D}(\widetilde{f})}(\rho)) \right) \cup \mathscr{FP}_{\mathscr{D}(\widetilde{f})}(\rho),$$

say $(x, y) \in \mathscr{FP}_{\mathscr{D}(\widetilde{f})}(\rho)$. Then $\exists (\sigma, \omega) \in \mathscr{D}(\widetilde{f})$ such that $\forall i \in \text{Supp}(\sigma), \pi_i(x, y) > 0, \forall j \in \text{Supp}(\omega), \rho_i(x, y) > 0\}$. Since $(\sigma, \omega) \in \mathscr{D}(\widetilde{f}), \widetilde{f}(\rho)(x, y) > 0$. Suppose

$$(x, y) \in (X \times X) \backslash \text{Symm}(\mathscr{FP}_{\mathscr{D}(\widetilde{f})}(\rho)).$$

Then

$$(y, x) \in (X \times X) \backslash \text{Symm}(\mathscr{FP}_{\mathscr{D}(\widetilde{f})}(\rho))$$

and so

$$\widetilde{f}_{\mathscr{D}(\widetilde{f})}(\rho)(y, x) = \widetilde{f}_{\mathscr{D}(\widetilde{f})}(\rho)(x, y) = 1.$$

However, this contradicts the assumption that $\pi_{\mathscr{D}(\widetilde{f})}(x, y) > 0$. Thus

$$(y, x) \in \text{Symm}(\mathscr{F}\mathscr{P}_{\mathscr{D}(\widetilde{f})}(\rho)) \backslash \mathscr{F}\mathscr{P}(_{\mathscr{D}(\widetilde{f})}(\rho)).$$

Hence it follows that $\widetilde{f}(\rho)(y, x) = 0$. Now suppose that $\widetilde{f}$ is neutral and monotonic. Let $x, y \in X$ and $\rho \in \mathscr{F}\mathscr{R}^n$. Suppose that $\pi(x, y) > 0$. Let $(\sigma, \omega) \in \mathscr{F}\mathscr{P}(N) \times \mathscr{F}\mathscr{P}(N)$, $\text{Supp}(\sigma) \subseteq \text{Supp}(\omega)$, be such that $\text{Supp}(\sigma) = S$ and $\text{Supp}(\omega) = W$, where $S = P(x, y; \rho)$ and $W = R(x, y; \rho)$. We wish to show $(\sigma, \omega) \in \mathscr{D}(\widetilde{f})$. Let $z, w \in X$. Let $\rho' \in \mathscr{F}\mathscr{R}^n$ be such that $\pi'_i(z, w) > 0 \Leftrightarrow i \in S$ and $\rho'_i(z, w) > 0 \Leftrightarrow i \in W$, i.e., $P(z, w; \rho') = S$ and $R(z, w; \rho') = W$. Since $\widetilde{f}$ is neutral, $\pi'(z, w) > 0$. Now let $\rho'' \in \mathscr{F}\mathscr{R}^n$ be such that $P(z, w; \rho') \subseteq P(z, w; \rho'')$ and $R(z, w; \rho') \subseteq R(z, w; \rho'')$. Since $\widetilde{f}$ is monotonic, $\pi''(z, w) > 0$. Thus we have that $\pi_i(z, w) > 0 \forall i \in S$ and $\rho_i(z, w) > 0 \forall i \in W$ implies $\pi(z, w) > 0$. Thus $(\sigma, \omega) \in \mathscr{D}(\widetilde{f})$ and so $\pi_{D(\widetilde{f})}(x, y) > 0$.

3.4 Single-Peaked Fuzzy Profiles

We now turn to our main results, a proof that Black's Median Voter Theorem holds with fuzzy social preferences (Theorem 3.43). We begin by considering single-peakedness. We then follow the lead in Austen-Smith and Banks (1999) and introduce the notion of a core by relating fuzzy $\widetilde{f}$-medians to single-peakedness. The core is a maximal set (a set of undominated alternatives) under some preference aggregation rule. Theorem 3.43 concludes that the core is located at the median when preferences of all political actors are single-peaked in single-dimensional policy space.

Definition 3.35 Suppose $|X| = n \geq 2$. Suppose Q is a strict ordering of X. Label X so that $\forall a_t \in X, a_{t+1}Qa_t, t = 1, 2, ..., n - 1$. Let $\pi \in \mathscr{F}\mathscr{R}$. Then π is called **single-peaked** on X with respect to Q if

$$\pi(a_t, a_{t+1}) > 0, \pi(a_{t+1}, a_{t+2}) > 0, ..., \pi(a_{n-1}, a_n) > 0 \text{ and}$$
$$\pi(a_t, a_{t-1}) > 0, \pi(a_{t-1}, a_{t-2}) > 0, ..., \pi(a_2, a_1) > 0.$$

Recall that π in Definition 3.35 is transitive.

Definition 3.36 (*Single-Peakedness*) Let $\rho \in \mathscr{F}\mathscr{R}^n$. Then ρ is said to be **single-peaked** on X if there exists a strict ordering Q of X such that for all $i \in N$, ρ_i is single-peaked on X with respect to Q.

Let $\mathscr{F}\mathscr{S}$ denote the set of all single-peaked fuzzy profiles.

Definition 3.37 Let $\rho \in \mathscr{F}\mathscr{R}^n$. Suppose ρ is single-peaked with respect to a strict ordering Q. Define x_i to be that element of X such that $\pi_i(x_i, y) > 0 \forall y \in X (x_i \neq y), i = 1, ..., n$. Define $\widetilde{L}^-, \widetilde{L}^+ : X \to \mathscr{F}\mathscr{P}(N)$ by $\forall z \in X, \forall i \in N$,

$$\widetilde{L}^-(z)(i) = \begin{cases} \pi_i(x_i, z) & \text{if } zQx_i, \\ 0 & \text{otherwise.} \end{cases}$$

$$\widetilde{L}^+(z)(i) = \begin{cases} \pi_i(x_i, z) & \text{if } x_iQz, \\ 0 & \text{otherwise.} \end{cases}$$

Definition 3.38 Let $\widetilde{f}$ be a fuzzy preference aggregation rule. Let $\rho \in \mathscr{FR}^n$. Suppose ρ is single-peaked with respect to a strict ordering Q. Then an element z of X is called an $\widetilde{f}$-**median** if $\text{Supp}(\widetilde{L}^-(z)) \notin \{\text{Supp}(\lambda) \mid \lambda \in \mathscr{L}(\widetilde{f})\}$ and $\text{Supp}(\widetilde{L}^+(z)) \notin \{\text{Supp}(\lambda) \mid \lambda \in \mathscr{L}(\widetilde{f})\}$.

Definition 3.39 Let $F : [0, 1]^2 \to [0, 1]$. $\forall \rho \in \mathscr{FR}^*(X), \forall \mu \in \mathscr{FP}^*(X), \forall x \in X$, define the fuzzy subset $M_F(\rho, \mu)$ of X by $\forall x \in X$,

$$M_F(\rho, \mu)(x) = \bigvee \{t \in [0, 1] \mid F(\rho(x, y), \mu(x)) \geq t \forall y \in \text{Supp}(\mu)\}.$$

The fuzzy subset $M_F(\rho, \mu)$ is called the **maximal fuzzy subset** of ρ and μ with respect to F.

Definition 3.40 Let $\widetilde{f}$ be a fuzzy preference aggregation rule. Let $F : [0, 1]^2 \to [0, 1]$. Define $\forall \rho \in \mathscr{FS}$ the fuzzy subset $\mu_{F\widetilde{f}}(\rho; Q)$ of X, or more briefly when the strict order Q for which ρ is single-peaked is understood, $\mu_{F\widetilde{f}}(\rho)$, by $\forall z \in X$,

$$\mu_{F\widetilde{f}}(\rho, Q)(z) = \begin{cases} \bigwedge \{F(\widetilde{f}(\rho)(z, y), 1) \mid y \in X\} & \text{if } z \text{ is an } \widetilde{f}\text{-median} \\ 0 & \text{otherwise.} \end{cases}$$

In Definition 3.40, $\mu_{F\widetilde{f}}(\rho)$ is the fuzzy subset of $\widetilde{f}$-medians given ρ with respect to F.

Definition 3.41 (*Fuzzy Core*) Let $F : [0, 1]^2 \to [0, 1]$. For all $\rho \in \mathscr{FR}^n$, define $C_{F\widetilde{f}}(\rho)(x) = M_F(\widetilde{f}(\rho), 1_X)(x)$ for all $x \in X$. Then $C_{F\widetilde{f}}(\rho)$ is called the **fuzzy core** of $\widetilde{f}$ at ρ with respect to F.

If $F = \wedge$, we use the notation $C_{\widetilde{f}}(\rho)$ for $C_{F\widetilde{f}}(\rho)$, $\mu_{\widetilde{f}}(\rho)(x)$ for $\mu_{F\widetilde{f}}(\rho)(x)$, and M for M_F.

Example 3.42 Let $F = \wedge$. Let $X = \{x, y, z\}$ and $N = \{1, 2, 3\}$. Let $\rho = (\rho_1, \rho_2, \rho_3) \in \mathscr{FR}^3$ be such that

$$\pi_1(x, y) = \tfrac{1}{8}, \quad \pi_1(y, z) = \tfrac{3}{4}, \quad \pi_1(x, z) = \tfrac{7}{8},$$

$$\pi_2(y, x) = \tfrac{1}{4}, \quad \pi_2(x, z) = \tfrac{5}{8}, \quad \pi_2(y, z) = \tfrac{7}{8},$$

$$\pi_3(z, y) = \tfrac{1}{2}, \quad \pi_3(y, x) = \tfrac{3}{4}, \quad \pi_3(z, x) = 1.$$

Let Q be such that $a_3 Q a_2 Q a_1$, where $a_3 = z$, $a_2 = y$, and $a_1 = x$. It follows that ρ is single-peaked with respect to Q, where for $\pi_1, t = 1$, for $\pi_2, t = 2$, and for

$\pi_3, t = 3$. Now $x_1 = x$, $x_2 = y$, and $x_3 = z$ since $\pi_1(x, w) > 0 \ \forall w \in X, w \neq x$, $\pi_2(y, w) > 0 \ \forall w \in X, w \neq y, \pi_3(z, w) > 0 \ \forall w \in X, w \neq z$. Hence

$$\widetilde{L}^-(x)(1) = 0 \text{ since not } xQx,$$
$$\widetilde{L}^-(x)(2) = 0 \text{ since not } xQy,$$
$$\widetilde{L}^-(x)(3) = 0 \text{ since not } xQz,$$
$$\widetilde{L}^-(y)(1) = \pi_1(x, y) \text{ since } yQx,$$
$$\widetilde{L}^-(y)(2) = 0 \text{ since not } yQy,$$
$$\widetilde{L}^-(y)(3) = 0 \text{ since not } yQz,$$
$$\widetilde{L}^-(z)(1) = \pi_1(x, z) \text{ since } zQx,$$
$$\widetilde{L}^-(z)(2) = \pi_2(y, z) \text{ since } zQy,$$
$$\widetilde{L}^-(z)(3) = 0 \text{ since not } zQz.$$

and

$$\widetilde{L}^+(x)(1) = 0 \text{ since not } xQx,$$
$$\widetilde{L}^+(x)(2) = \pi_2(y, x) \text{ since } yQx,$$
$$\widetilde{L}^+(x)(3) = \pi_3(z, x) \text{ since } zQx,$$
$$\widetilde{L}^+(y)(1) = 0 \text{ since not } xQy,$$
$$\widetilde{L}^+(y)(2) = 0 \text{ since not } yQy,$$
$$\widetilde{L}^+(y)(3) = \pi_3(z, y) \text{ since } zQy,$$
$$\widetilde{L}^+(z)(1) = 0 \text{ since not } xQz,$$
$$\widetilde{L}^+(z)(2) = 0 \text{ since not } yQz,$$
$$\widetilde{L}^+(z)(3) = 0 \text{ since not } zQz.$$

Let $\widetilde{f}$ be a simple majority rule. Let $\lambda \in \mathscr{L}(\widetilde{f})$. Then $\forall \rho \in \mathscr{F}\mathscr{R}^n$, $\pi_i(x, y) > 0 \ \forall i \in$ Supp$(\lambda) \Rightarrow \pi(x, y) > 0$. Thus

$$\left\{ \text{Supp}(\lambda) \mid \lambda \in \mathscr{L}(\widetilde{f}) \right\} = \{\{1, 2\}, \{1, 3\}, \{2, 3\}, \{1, 2, 3\}\} \ .$$

Since

$$\text{Supp}(\widetilde{L}^-(x)) = \emptyset, \text{Supp}(\widetilde{L}^-(y)) = \{1\}, \text{Supp}(\widetilde{L}^-(z)) = \{1, 2\},$$
$$\text{Supp}(\widetilde{L}^+(x)) = \{2, 3\}, \text{Supp}(\widetilde{L}^+(y)) = \{3\}, \text{Supp}(\widetilde{L}^+(z)) = \emptyset,$$

we have that Supp$(\widetilde{L}^-(y)) \notin \{\text{Supp}(\lambda) \mid \in \mathscr{L}(\widetilde{f})\}$ and Supp$(\widetilde{L}^+(y)) \notin \{\text{Supp}(\lambda) \mid \lambda \in \mathscr{L}(\widetilde{f})\}$. Hence y is an $\widetilde{f}$-median. Thus $\mu_{\widetilde{f}}(\rho; Q)(y) = \widetilde{f}(\rho)(y, x) \wedge \widetilde{f}(\rho)(y, z)$. Now $M(\widetilde{f}(\rho), 1_X)(y) = \bigvee\{t \in [0, 1] \mid \widetilde{f}(\rho)(y, w) \geq t \forall w \in X\} = \widetilde{f}(\rho)(y, x) \wedge \widetilde{f}(\rho)(y, z)$. It follows that

$$\widetilde{f}(\rho)(y, x) = \pi_2(y, x) \vee \pi_3(y, x)$$
$$= \frac{1}{4} \vee \frac{3}{4}$$
$$= \frac{3}{4},$$
$$\widetilde{f}(\rho)(y, z) = \pi_1(y, z) \vee \pi_2(y, z)$$
$$= \frac{3}{4} \vee \frac{7}{8}$$
$$= \frac{7}{8}.$$

Thus $\widetilde{f}(\rho)(y, x) \wedge \widetilde{f}(\rho)(y, z) = \frac{3}{4}$.

We now present our main conclusion.

Theorem 3.43 *Let $\widetilde{f}$ be a fuzzy aggregation rule. Assume that $F : [0, 1]^2 \to [0, 1]$ is such that $F(0, 1) = 0$. If $\widetilde{f}$ is a partial fuzzy simple rule, then for any $\rho \in FS$, $C_{F\widetilde{f}}(\rho)(x) = \mu_{F\widetilde{f}}(\rho, Q)(x)$ for all $x \in X$, where Q is the strict order for which ρ is single-peaked.*

Proof Let $\rho \in \mathscr{FS}$ and $z \in X$. Then

$$M_F(\widetilde{f}(\rho), 1_X)(z) = \bigvee \left\{ t \in [0, 1] \mid F(\widetilde{f}(\rho)(z, y), 1_X(x)) \geq t \; \forall y \in \mathrm{Supp}(1_X) \right\}$$
$$= \bigvee \left\{ t \in [0, 1] \mid F\left(\widetilde{f}(\rho)(z, y), 1\right) \geq t \; \forall y \in X \right\}$$
$$= \bigwedge \left\{ F\left(\widetilde{f}(z, y)(\rho), 1\right) \mid y \in X \right\}.$$

Hence if z is an $\widetilde{f}$-median, $C_{F\widetilde{f}}(\rho)(z) = \mu_{F\widetilde{f}}(\rho)(z)$. That is, $M_F(\widetilde{f}(\rho), 1_X) = \mu_{F\widetilde{f}}(\rho)$ on $\mathrm{Supp}(\mu_{F\widetilde{f}}(\rho))$. Hence it suffices to show

$$\mathrm{Supp}(M_F(\widetilde{f}(\rho), 1_X)) = \mathrm{Supp}(\mu_{F\widetilde{f}}(\rho)).$$

Let $x \in \mathrm{Supp}(M_F(\widetilde{f}(\rho), 1_X))$. Then

$$F(\widetilde{f}(\rho)(x, y), 1_X(x)) = F(\widetilde{f}(\rho)(x, y), 1) > 0$$

for all $y \in X$. Suppose $\mu_{F\widetilde{f}}(\rho) = 0$, i.e., $x \notin \mathrm{Supp}(\mu_{F\widetilde{f}}(\rho))$. Let Q be a strict ordering of X with respect to which ρ is single-peaked. Since X is finite, we are free to label X so that $a_{t+1} Q a_t$ for all $t \geq 1$ (and $t \leq |X|$). Then $x = a_t$ for some $t \geq 1$ and x is not an $\widetilde{f}$-median since $x \notin \mathrm{Supp}(\mu_{F\widetilde{f}}(\rho))$. Hence either

$$\mathrm{Supp}(\widetilde{L}^-(x)) \in \left\{ \mathrm{Supp}(\lambda) \mid \lambda \in \mathscr{L}(\widetilde{f}) \right\}$$

or

$$\mathrm{Supp}(\widetilde{L}^+(x)) \in \{\mathrm{Supp}(\lambda) \mid \lambda \in \mathscr{L}(\widetilde{f})\} \ .$$

Assume the former. Then $\mathrm{Supp}(\widetilde{L}^-(x)) = \mathrm{Supp}(\lambda)$ for some $\lambda \in \mathscr{L}(\widetilde{f})$. Thus $\forall u, v \in X$, $\pi_i(u, v) > 0 \ \forall i \in \widetilde{L}^-(x)$ implies $\pi(u, v) > 0$. Since ρ is single-peaked, $\pi_i(a_{t-1}, a_t) > 0 \forall i \in \mathrm{Supp}(\widetilde{L}^-(x))$. Hence since $\mathscr{L}(\widetilde{f})$ is decisive, $\widetilde{f}(\rho)(a_{t-1}, a_t) > 0$, i. e., $\widetilde{f}(\rho)(a_{t-1}, x) > 0$. But then $\widetilde{f}(\rho)(x, a_{t-1}) = 0$ and so from the assumption that $F(0, 1) = 0$, we have $F(\widetilde{f}(\rho)(x, a_{t-1}), 1) = 0$, a contradiction. It therefore follows that $x \in \mathrm{Supp}(\mu_{F\widetilde{f}}(\rho))$. Hence

$$\mathrm{Supp}(M_F(\widetilde{f}(\rho), 1_X)) \subseteq \mathrm{Supp}(\mu_{F\widetilde{f}}(\rho)).$$

Since $\mathrm{Supp}(\mu_{F\widetilde{f}}(\rho)) \subseteq \mathrm{Supp}(M_F(\widetilde{f}(\rho), 1_X))$ we have that $\mathrm{Supp}(M_F(\widetilde{f}(\rho), 1_X)) = \mathrm{Supp}(\mu_{F\widetilde{f}}(\rho))$. Hence $M_F(\widetilde{f}(\rho), 1_X) = \mu_{F\widetilde{f}}(\rho)$. Now assume that $\mathrm{Supp}(\widetilde{L}^+(x)) \in \{\mathrm{Supp}(\lambda) \mid \lambda \in \mathscr{L}(\widetilde{f})\}$. Then $\mathrm{Supp}(\widetilde{L}^+(x)) = \mathrm{Supp}(\lambda)$ for some $\lambda \in \mathscr{L}(\widetilde{f})$. Thus $\forall u, v \in X$, $\pi_i(u, v) > 0 \forall i \in \widetilde{L}^+(x)$ gives $\pi(u, v) > 0$. Since ρ is single-peaked, $\pi_i(a_{t+1}, a_t) > 0 \forall i \in \mathrm{Supp}(\widetilde{L}^+(x))$. Hence since $\mathscr{L}(\widetilde{f})$ is decisive, $\widetilde{f}(\rho)(a_{t+1}, a_t) > 0$, i. e., $\widetilde{f}(\rho)(a_{t+1}, x) > 0$. But then as before, $F(\widetilde{f}(\rho)(x, a_{t+1}) = 0$, a contradiction. Thus $x \in \mathrm{Supp}(\mu_{F\widetilde{f}}(\rho))$. Hence we have that $\mathrm{Supp}(M_F(\widetilde{f}(\rho), 1_X)) \subseteq \mathrm{Supp}(\mu_{F\widetilde{f}}(\rho))$. Since

$$\mathrm{Supp}(\mu_{F, \widetilde{f}}(\rho)) \subseteq \mathrm{Supp}(M_F(\widetilde{f}(\rho), 1_X)) \ ,$$

we have

$$\mathrm{Supp}(M_F(\widetilde{f}(\rho), 1_X)) = \mathrm{Supp}(\mu_{F\widetilde{f}}(\rho)).$$

We conclude that

$$M_F(\widetilde{f}(\rho), 1_X) = \mu_{F\widetilde{f}}(\rho).$$

Corollary 3.44 *Let* $F = \wedge$. *Let* $\widetilde{f}$ *be a fuzzy aggregation rule. If* $\widetilde{f}$ *is a partial fuzzy simple rule, then*

$$\forall \rho \in \mathscr{FS}, C_{\widetilde{f}}(\rho) = \mu_{\widetilde{f}}(\rho).$$

Theorem 3.43 demonstrates that when preferences are single-peaked, a simple rule produces a maximal set, or core in single-dimensional public choice models. We turn in the next chapter to testing a fuzzy single-dimensional fuzzy public choice model. We begin by applying Black's Median Voter Theorem.

References

Austen-Smith, D., Banks, J.: Positive Political Theory I. The University of Michigan Press, Ann Arbor (1999)

Black, D.: The Theory of Committees and Elections. Cambridge University Press, Cambridge (1958)

Downs, A.: An Economic Theory of Democracy. Harper and Row Publishers, New York (1957)

Gibilisco, M.B., Mordeson, J.N., Clark, T.D.: Fuzzy black's median voter theorem: examining the structure of fuzzy rules and strict preference. New Math. Nat. Comput. **8**, 195–217 (2012)

Kiewit, D., McCubbins, M.D.: Presidentia influence on congressional appropriation decisions. Am. J. Polit. Sci. **32**, 713–736 (1988)

McKelvey, R.D.: Intransitives in multidimensional voting models and some implications for agenda control. J. Econ. Theor. **12**(3), 472–482 (1976)

Mordeson, J.N., Nielsen, L., Clark, T.D.: Single-peaked fuzzy preferences in one-dimensional models: does black's median voter theorem hold? New Math. Nat. Comput. **6**, 1–16 (2010)

Chapter 4
Predicting the Outcome of the Government Formation Process: Fuzzy Single-Dimensional Models

Abstract We are now in a position to present a one-dimensional fuzzy public choice model designed to predict the outcome of the government formation process in parliamentary systems. Such a model allows us to represent flexibility in actors' preferences and predict when those actors may make allowances for minor policy shifts as well as when they may prefer major policy shifts. This is because the fuzzy public choice model allows for broad areas of indifference in actor's preference profiles. Moreover, a fuzzy model is more likely to predict stable outcomes by avoiding the intransitivity problem that plagues traditional models. We present two approaches to such a model. The first makes use of the fuzzy maximal set; the second makes use of the fuzzy Pareto set. We test both models using fuzzy preferences derived from the Comparative Manifesto Project (CMP) data.

4.1 An Overview of the Comparative Manifesto Project Data

In this chapter, we extract fuzzy preference measures from CMP data using the bootstrap procedure designed by Benoit et al. to calculate error in the CMP data as we previously discussed. We apply these preference measures to a one-dimensional fuzzy choice model designed to predict the outcome of the government formation process. We then compare the predictions of the fuzzy government formation model with data about the actual governing coalitions formed following each election. The results of the test give us a sense of the accuracy of our model and the preference measures we use to ground it empirically.

We begin by giving careful consideration to the Comparative Manifesto Project (CMP) data. Out of the 54 countries and 529 election years represented in the CMP data, we select all multiparty European democracies with either a parliamentary or premier-presidential regime. Shugart and Carey (1992) define parliamentary and premier-presidential regimes as those in which the government, or cabinet, is

P. C. Casey et al., *Fuzzy Social Choice Models*, Studies in Fuzziness
and Soft Computing 318, DOI: 10.1007/978-3-319-08248-6_4,
© Springer International Publishing Switzerland 2014

accountable to the legislature and not a directly elected president. This selects out countries like Mexico and the United States that have presidential democratic systems in which the cabinet is accountable to the president only, or president-parliamentary systems such as in Russia and Ukraine where the cabinet has mixed accountability to the president and parliament. We also select out any countries with two-party systems or dominant-party systems in which the government is likely to be formed by a single party, like Great Britain or Japan, since prediction of government formation in these cases is arbitrary. BLM (2009) also select out all cases in which 99.99 % of the quasi-sentences in the CMP data are recorded as "uncoded" (per uncode). These include Sweden from 1948 to 1982 and all Norwegian cases.

The resulting set of data includes 25 European countries and 265 election years from 1945 to 2005. Eleven of the remaining countries began holding democratic elections in the 1940s, three are southern European countries (Spain, Greece, and Portugal) that established democratic elections in the 1970s, and eleven are Central and Eastern European (CEE) countries that established democratic elections in the 1990s following the collapse of the Soviet Union.

We use data from Müller and Strøm (2000) to determine the actual, real-world results of the government formation process following each election in the CMP data base. We consider governments formed immediately following an election only, eliminating changes in the cabinet between electoral periods. Data on actual government formation was missing for 30 of the 265 cases, leaving 235 country-years against which to test the models. Eighteen of the missing cases of government formation are cases since 1999. Three of the missing cases of government formation are from the French Fourth Republic, which was not recorded by Müller and Strøm. The remaining nine missing cases are the outcomes of government formation following some of the earliest elections in a country; one following an election in the 1940s (Italy), two from the 1970s (Portugal and Greece), and six from CEE countries in the early 1990s.

The data span seven decades with 16 elections (6.81 %) taking place in the 1940s, 31 (13.19 %) in the 1950s, 28 (11.91 %) in the 1960s, 40 (17.02 %) in the 1970s, 45 (19.15 %) in the 1980s, 65 (27.66 %) in the 1990s, and 9 (3.83 %) since 2000. Of the governments formed immediately following the 235 elections in our dataset, 44 (18.72 %) were single-party minority governments, 27 (11.49 %) were minority coalitions, 23 (10.21 %) were single-party majority governments, 89 (37.45 %) were minimum-winning coalitions, and 52 (22.13 %) were supermajority governments.[1]

Of the 3018 total policy positions identified by the CMP, 1511 are captured in our dataset. While all 3018 were converted into fuzzy policy positions, only 1511 cases were in European multiparty parliamentary or premier-presidential democracies. These 1511 policy positions were distributed across the *rile* spectrum in the original CMP data. The right-most *rile* position was that of the Icelandic Progressive Party (F) in 1974 at a position of 82.20 on the *rile* dimension. The left-most position was that

---

[1] Percentages rounded to the nearest hundredth.

of Luxembourg's Communist Party (PCL/KPL) in 1979 at −74.30 on the *rile* issue dimension. Belgium's Flemish Christian People's Union (FCPU) in 1965 was at the median *rile* position for all 1511 cases at −1.62, just slightly left of center.[2]

Although they are similar, the median of parties' fuzzy policy positions (fuzzy median) differs slightly from their *rile* position determined by Benoit et al. (2009). This is to be expected, as each party's preference profile is based on the density function of the distribution of points from the bootstrap procedure, and not on the parties' *rile* position. In 1948, the fuzzy median of Finland's Agrarian Union was the median position for all 1511 cases, slightly left of center at −1.88. This is notable as the party's *rile* position was not the median, slightly left of center at −2.33. In fact, nineteen other parties were closer to the fuzzy median than the 1948 Finnish Agrarian Union. Similarly, the right-most fuzzy median, held by Finland's 1970 Christian Union (SKL) at 90.74, differs from its *rile* position 78.85, and is further right than Iceland's 1974 Progressive Party, which has the second right-most fuzzy median at 90.35. The left-most fuzzy median also differs from the left-most *rile* position. Denmark's 1971 Left Socialist Party (VS) is the left-most median at −100 (*rile* = −12.10). However, the Danish VS controlled no parliamentary seats in 1971, and the party with the left-most fuzzy median to be elected to legislature is Denmark's Socialist People's Party (SF) at −82.56 (*rile* = −68.10). Luxembourg's 1979 PCL/KPL, the left-most *rile* position, has the fifth left-most fuzzy median.

While the median of a fuzzy number gives us some sense of its position, it tells us very little about a party's fuzzy preferences, since fuzzy preferences are not a single point, but a subset of policy space. This is extremely different from the Euclidean model in which only position matters. In the fuzzy model, we are concerned both with the position of the fuzzy number and its shape. When we say we are concerned with the shape of a fuzzy number, we are really interested in its range; the amount of area within the policy space it contains. This can give us a sense of how likely a fuzzy preference profile is to intersect the preference profile of another player. Therefore, the range of the support and the position of its bounds are of particular interest to us.

The supports of fuzzy numbers vary greatly in our dataset of 1511 preference profiles. The political party with the fuzzy preference profile with the largest support is Denmark's 1950 Radical Party (RV). The 1950 Danish RV's fuzzy preference profile extends from a left-bound at −59.49 and a right-bound at 60.69, for a total range of 120.18 along the *rile* dimension. The minimum range of the support of a fuzzy preference profile is that of Greece's 1996 New Democracy Party (ND), whose support extends from −12.59 to −10.03 for a total range of just 2.55 along the *rile* dimension. The median range of the fuzzy support for all 1511 fuzzy preference profiles is that of Italy's 1972 Republican Party (PRI) with a left-bound at −17.26 and a right-bound at 7.35 for a total range of 24.61 along the *rile* dimension. This demonstrates that political actors' perception of shifts in policy space varies significantly, suggesting that Euclidean assumptions are insufficient for characterizing player's preference profiles.

---

[2] All numbers reported for rile, fuzzy median, ranges of the support of fuzzy numbers, and other descriptive statistics are rounded to the nearest hundredths place.

The relationship between the size of the core of a player's fuzzy preferences and the player's support is also an indicator of a player's "tolerance" of policy change. As mentioned above, the core of a fuzzy number is always a subset of its support. However there can be variation in the difference between the range of the support and the range of the core. For example, the party with the largest core in our dataset is Latvia's 1993. For the Fatherland and Freedom (TUB) with a core range of 24.21 on the *rile* dimension, about equal to the median range of the support in the dataset. However, the range of the support of the party's fuzzy preference profile is 83.62, significantly smaller than the maximum, that of the 1953 Danish RV. However, since TUB's core range is the greatest, RV's core range is smaller while its support range is larger, demonstrating a difference in how the parties perceive shifts in policy space. It seems that, while TUB views a wider range of policies as ideal than does RV, RV is more tolerant of policy shifts away from its ideal than TUB. Similarly, Greece's 1996 Panhellenic Socialist Movement (PASOK) has the smallest core range at 0.59 along the *rile* dimension but not the smallest support range. The median core range for a fuzzy preference profile in our dataset of 1511 is Iceland's 1999 Liberal Party (FF) at 5.08 along the *rile* dimension.

Differences in the support ranges of parties' fuzzy preference profiles appear to be correlated with the decade of the election and the party's position on the *rile* dimension. The average range of parties' fuzzy preference profiles appears to decrease over each consecutive decade. The average range of parties' fuzzy preferences in the 1940s was about 39.70. By the 1980s, the average range of parties' fuzzy preferences had decreased to about 21.50. Thus, their fuzzy preference profiles become less "tolerant" of policy shifts; the party is more likely to experience decreases in utility with shifts away from its ideal point.

Contrary to this pattern, in the 1990s the average range of parties' preferences increased again to about 24.50, later decreasing to 22.56 in the 2000s. However, when the 11 CEE countries are removed, whose democracies began in the 1990s, the pattern is re-established, with the average range falling to 20.81. The fact that the younger CEE democracies demonstrate fuzzy preference profiles with broader support ranges further justifies the explanation that younger parties' preference profiles demonstrate more tolerance for policy shifts. Furthermore, the 1990s was a decade in which many European democracies saw the emergence of new political parties. The introduction of these new parties could explain the very small decrease in the average range of parties' fuzzy preferences from the 1980s to the 1990s. New parties may maintain broader, more tolerant preference profiles than older, more established parties.

The size of the range of the support of a party's fuzzy preference profile also appears to be correlated with the position of the party's fuzzy median on the *rile* dimension, though not with the party's *rile* position. This makes sense since parties' *rile* positions and fuzzy medians are only loosely associated. The average width of the support of a party's fuzzy preference profile almost takes the shape of a reverse bell curve in relation to the *rile* dimension. Parties with positions at the extremes of the *rile* dimension tend to have much wider supports while parties toward the center have much narrower fuzzy preference profiles. The average width of the support of parties' fuzzy preference profiles with a median f such that $-100 < f < -75$

was about 45.86. However, as the fuzzy median of parties' preference profiles shifts towards $-75 < f < -50$ and $-50 < f < -25$, the average width of parties' preference profiles decreases to 29.77 and 25.95, respectively. At $-25 < f < 0$, the average width of the support of a party's fuzzy preference profile decreases to about 23.17. The average width of the support then increases for parties with a fuzzy median at $0 < f < 25$ to about 27.72 and then to 35.21 and 41.06 for parties with a fuzzy median at $25 < f < 50$ and $50 < f < 75$, respectively. This indicates that parties with fuzzy preference profiles on the margins of the *rile* dimension may be more tolerant of policy shifts away from their ideal position than those towards the center. However, the pattern is broken for parties with a fuzzy median at $75 < f < 100$, where the width of the support of parties' fuzzy preference profiles decreases suddenly to 29.91.

Parties' fuzzy preferences broaden with the diversity of issues mentioned within their political texts. If a party mentions almost exclusively issues on a single wing of the *rile* dimension, then its fuzzy preference profile is more likely to be narrow and almost exclusively on that wing of the rile dimension. However, if the party mentions policies on both wings of the *rile* dimension or issues that are not on the *rile* dimension, the parties fuzzy preference profile will tend to be broader and may straddle both sides of the dimension. This suggests that parties whose fuzzy preference profiles narrow over time are narrowing the scope of policies they emphasize in their political texts. If this is true, then we may propose that parties narrow the range of policies they mention over time and become more specific and clearer about where the position themselves along the *rile* dimension. Similarly, we may suppose that parties toward the extremes of the *rile* dimension express support for a broader range of issues, or are especially concerned with another non-*rile* issue, because of the breadth of their fuzzy preference profiles. If this is true, then parties towards the median of the policy space are not expressing support for policies on either end of the *rile* dimension, but are making very direct, explicit reference to their position on the dimension. If this is the case, then we may presume that more moderate parties maintain narrower, more specific policy platforms than parties whose positions are more extreme. The exception would be parties on the far-right of the *rile* dimension who maintain very narrow policy platforms specifically identifying themselves as far right-wing.

The breadth and placement of parties' fuzzy preference profiles is important because it can determine the likely outcomes of our fuzzy government formation model. If parties' fuzzy preference profiles are broader and more concentrated, then intersections between their preferences will be more likely. However, if their preference profiles are narrower and more disperse, then intersections will be less prevalent. The parties' preference intersect, the more potential legislative coalitions are possible and the more likely the fuzzy model will predict a legislative coalition that forms the real-world government. Less intersection between parties' fuzzy preference profiles implies fewer possible legislative coalitions, decreasing the likelihood that the governing coalition will be captured by the model, but increasing the precision of accurate predictions.

From our review of the data, more moderate policy positions, which tend to be more clustered, are narrower, while more extreme positions, which tend to be isolated, are broader. This may increase the likelihood of centrally-located parties having the opportunity to form legislative coalitions with a diverse group of partners either at more moderate policy positions towards the center of the *rile* dimension or less moderate positions towards the dimension's extremes. What seems clear is that a model making use of fuzzy preferences has the distinct possibility of predicting a diverse set of potential legislative coalitions, in contrast to the Euclidean models whose predictions will always be biased towards the median. This may increase the likelihood of fuzzy making correct predictions of government formation. It is to this question that we now turn.

4.2 Traditional Models of Government Formation

A government forms whenever the previous government is replaced by a new government. We are particularly concerned with government formation in parliamentary and premier-presidential democratic systems[3] in which the government is accountable to the legislature by way of a vote of confidence. Governments in these systems need to retain the support of at least an absolute majority of the legislature in order to stay in office. A government is formed by a coalition of one or more parties that is capable of achieving the confidence of at least an absolute majority of the legislature. We differentiate between a governing coalition and a legislative coalition. A legislative coalition is the set of parties that support the government in a vote of confidence, while a governing coalition is the set of parties actually holding office in the government.

In predicting government formation, we are naturally concerned with the formation of the governing coalition. A governing coalition can take a number of forms. The most common is the minimum-winning coalition, in which the party or parties in the governing coalition control a majority of the seats in the legislature. The second most common is the minority governing coalition, where the party or parties in the governing coalition control less than a majority of seats in the legislature and therefore must rely on the support of other parties in a legislative coalition. The third type of coalition is the supermajority coalition, in which one or more parties in the coalition bring seats to the governing coalition which exceed the number of seats necessary to form a majority in the legislature.

Government formation has been a common application for formal models of public choice. A number of models have been proposed under Euclidean assumptions, in both single-dimensional and multi-dimensional space. If we consider the different models proposed by scholars to predict government formation on multiple policy dimensions, we can categorize the major models in the literature into two groups: models that assume ministerial dominance, and those that assume that parties manage delegated ministers through a variety of control mechanisms.

---

[3] As defined by Shugart and Carey (1992).

The first, the ministerial-dominance model, (Laver and Shepsle 1990, 1996) assumes that parties make policy trade-offs when they delegate ministers to specific policy areas. By making these trade-offs, parties sacrifice control and therefore implementation of their preferences in a certain policy area in return for policy control in another. Ministers are unimpeded while pursuing their own policy preferences or, at least, the preferences of their party. The jurisdiction of a minister over his own policy area implies that the minister can move policy to whatever position he wants without punishment from other parties in the coalition.

The second category of multidimensional government formation models assumes that the members of the governing coalition manage delegated ministers through a number of control mechanisms, such as committees and junior ministers (Theis 2001; Martin and Vanberg 2004). Under the managed-delegation model, members of a coalition agree upon a policy position for the newly-formed government, and it is the role of the delegated ministers to implement the agreed-upon policy. Coalition members use a number of control mechanisms to ensure that ministers delegated to certain policy areas do not stray from the status quo. This model makes ministers implementers of government policy. Any policy change, however slight from that agreed upon by the coalition, will result in a reprimand at the least and possibly dissolution of the government (Tsebelis 2002). Therefore, this model predicts a government in which the minister has no independent policy control in his area of delegation, effectively eliminating any policy incentives for holding office.

Clearly, the outcomes of the assumptions of either of these models are unsatisfactory. The assumptions of the ministerial-dominance model force us to accept that ministers can pursue whatever policies they prefer regardless of the preferences of his party's coalition partners. On the other hand, the managed-delegation model predicts policy-rigid governments forever on the precipice of dissolution. Neither of these outcomes seems comparable to our actual experience of coalition governments in developed democracies.

Moreover, many conventional one-dimensional models focus on the fact that parties must form simple majority legislative coalitions in order to win a vote of confidence and that they form these coalitions based on their relative policy positions. Under these assumptions, there is no motivation for governments to form coalitions that include more parties than are necessary to achieve a simple majority. As demonstrated by Black's Median Voter Theorem, the Euclidean distance assumption induces a median tendency in conventional models because each player attempts to achieve the policy closest to his ideal point. This leads to a kind of tug-of-war, which results in policy being implemented at the median player's position. Thus, single-dimensional conventional models tend to over-predict single-party minority governments falling at the median. Only under a skewed distribution of legislative weights is a governing coalition of multiple parties predicted. Furthermore, the conventional approach has difficulty explaining the formation of supermajority governments. While we might make the additional assumption that parties are likely to form supermajority coalitions in order to respond to some qualified majority requirement, such as overriding a presidential veto, such assumptions are contextually bound and therefore are not terribly generalizable.

4.3 A Fuzzy Model of Government Formation

A fuzzy government formation model offers an alternative to the median tendencies of the Euclidean model. In what follows, we present two fuzzy government formation models. The first makes use of the fuzzy maximal set to predict the outcome of the government formation process; the second makes use of the fuzzy Pareto set. Both models assume that any government must have the support of at least an absolute majority of the parliament in order to achieve and retain office. Therefore, the goal of any potential government is to form at a policy position supported by a legislative coalition controlling at least 50 % plus one seats of the parliament. Both models also assume that political parties are unitary and disciplined. In other words, the preferences of a party's members are identical to the preferences of the party, or at least individual members do not defect from the party position when it comes to collective decisions. Finally, both models assume that parties are purely policy-motivated rather than office-, vote-, or mixed-motivated. Under these assumptions, parties only make decisions based upon which policy best satisfies their preference utility and do not take any other motivation into consideration, such as office benefits or reelection.

Our fuzzy models require the use of a decision rule to determine which alternatives constitute the maximal set or Pareto set. The application of a decision rule somewhat reduces the usefulness of the fuzzy preference profile that was discussed in chapter two as it requires players determine whether xPy or xIy in a manner synonymous to the crisp model. This can be partially addressed by using a fuzzy preference relation, which permits a player to prefer, or be indifferent to, a set of alternatives to a varying degree. Models of government formation that include thick indifference in actors' preferences significantly reduce the likelihood of intransitivity in collective preferences.[4] However, conventional public choice models incorporating indifference are cumbersome and do not lend themselves to empirical research. Fuzzy set theory provides a simpler method for modeling indifference. Fuzzy set theory differs from traditional set theory because it asks to what degree is an element in a set instead of is an element in or out of a set. While the set inclusion level in traditional set theory must be 1 or 0, either in or out of the the the set, fuzzy set theory uses a characteristic function, which we call σ_i, to map the universe X into the interval $[0, 1]$, formally written as $\sigma_i : X \rightarrow [0, 1], \forall i \varepsilon N$. In policy terms, $\sigma_i(x)$ refers to the degree to which x is an ideal policy of individual i, where 1 is an ideal policy and 0 is a completely non-ideal policy.

As we discussed briefly in chapter two, there are several possible ways to conceptualize σ_i over a policy dimension. Figure 4.1 demonstrates a natural first extension of fuzzy sets to a one-dimensional public choice model, with fuzzy preference profile with an ideal position at x_i, where $\sigma_i(x_0) = 1$. While this depiction may seem identical to the traditional single-peaked profile, the fuzzy profile introduces the concept of a policy horizon, past which, an actor is unwilling to accept any policy. Because

---

[4] See Clark et al. (2008) for how thick indifference reduces majority cycling without including institutional restraints.

Fig. 4.1 Simple fuzzy preference profile

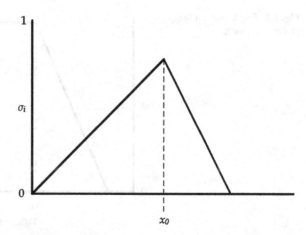

the actor views some policy as completely not ideal ($\sigma_i = 0$), the actor will consider any alternative in this range as unacceptable, effectively limiting the number
of acceptable alternatives.[5] The actor is completely indifferent in the policy area
$\sigma_i = 0$; it views all alternatives as equally abhorrent. In contrast, traditional single-
peaked models depict actors who are willing to accept any policy alternative if its
position is closer to their ideal point than another proposed alternative. Figure 4.1
also introduces the concept in fuzzy set theory called the support. The support of σ_i is
the group of alternatives such that $'\sigma_i > 0$. If an actor is to support any alternative, it
must lie within the actor's support (Clark et al. 2008). By acknowledging that actors
can only compromise to a certain extent, fuzzy set theory essentially restricts policy
space.

Indifference has the potential to induce stability because actors view dimension
spaces as discrete areas rather than a collection of infinite alternatives. A common
type of indifference in the spatial literature is single-plateau preferences, where actors
have an ideal policy range rather than a ideal point (Ching and Serizawa 1998; Masso
and Neme 2001). Figure 4.2 demonstrates a fuzzy, single-plateau preference profile.
The support is the same as Fig. 4.1, but Fig. 4.2 has another area of indifference.
The "flat-top", between x_1 and x_2, represents the core in fuzzy mathematics, where
$\sigma_i = 1$. The core refers to an actor's ideal range. Again the actor is completely
indifferent to all policy that lie within the core because all possible positions within
the core have the same σ-level.

The profile in Fig. 4.2 shows an actor with two different areas of indifference
(the core and outside of the support) but with Euclidean preferences connecting
the two. The actor is still able to discern minute shifts in policy between the interval
$\sigma_i = (0, 1)$, which poses two problems. First, there is a greater chance of intransitivity
because actors still possess some degree of Euclidean preferences. Second and more

---

[5] This idea of a policy horizon is also present in empirical research on government formation.
(Warwick 2005a,b) demonstrates political parties have policy horizons, past which they are unable
to accept any compromise on policy.

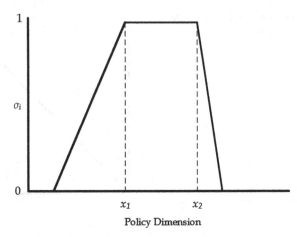

Fig. 4.2 Single-plateau fuzzy preference profile

importantly, an actor has significant areas of policy indifference, but at the same time possesses Euclidean preferences. Figure 4.2 then presents two competing views of human behavior, where one can be simultaneously indifferent over a broad policy range but also be able to detect small changes in policy in another range. Thus, a discretized characteristic function is more appropriate.

Fuzzy mathematics allows us to construct such a function. If σ_i is allowed to range continuously between 0 and 1, then a player's preferences are essentially Euclidian. If, however, σ is restricted to a discrete set, then a player is said to possess discretized preferences, or "thick- indifference".[6] Figure 4.3 presents such a profile. In this example, σ_i is an element of $0, 0.25, 0.5, 0.75, 1$. An actor is completely indifferent over all alternatives at a specific preference level because all alternatives have an equivalent σ_i. While researchers can choose any numbers of preference levels, a Likert-esque division has a natural appeal. A preference level of 1 can be interpreted as perfectly ideal, a level of 0.75 as almost ideal, a level of 0.5 as neither ideal nor not ideal, a level of 0.25 as not ideal and a level of 0 as completely not ideal.

In the models that we present, the proximity of players' fuzzy preferences is only important insofar as it determines the set inclusion value of the intersection of players' preferences. Players whose policy positions are located close to one another are more likely to intersect at higher α-levels. However, proximity neither guarantees nor precludes intersection. It is conceivable that two players with relatively close policy positions will intersect at low α-levels or not at all. The goal then in determining the outcomes of government formation with a fuzzy model is to identify the intersections between players' fuzzy preferences.

Our fuzzy models of the government formation process predict outcomes at the intersection of players' preferences regardless of their relative positions. Thus, collective decisions can occur anywhere in a single-dimensional policy space and

---

[6] Formally, an actor i possesses thick indifference if for $a, b \in X$, $a \neq b \nRightarrow a P_i b$ or $b P_i a$ where P_i is the strict preference relation for player i.

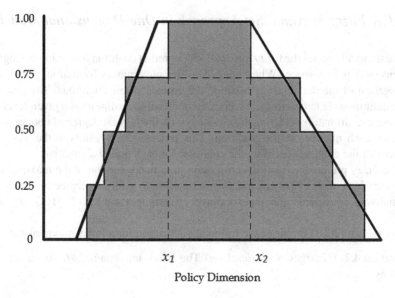

Fig. 4.3 Discrete fuzzy preference profile

can include any set of parties as long as players' preferences intersect and they have enough legislative weight to achieve their policy goals. This allows the fuzzy public choice model to predict government formation at other positions in single-dimensional policy space besides the median. Furthermore, the fuzzy model can predict a number of potential governing coalitions of any size in single-dimensional space. The number of such coalitions is only limited by the bounds of individual players' discrete α-levels.

Moreover, as we have discussed previously, rather than requiring us to assume that political actors have precise policy positions that they can only communicate with error, the use of fuzzy preferences in our models allows us to conceive of actors' preferences as vague and communicated accurately. It also allows us to shed the assumption that actors perceive shifts in utility in infinitely precise increments at the same granularity across and infinite policy space. Finally, fuzzy preferences permit our models to explain seemingly detrimental shifts in policy on one dimension in favor of gains on another dimension. In Euclidean public choice models, a political actor will only give way to a loss in utility on one dimension if it entails an overall gain in utility in multidimensional space. In contrast, a fuzzy public choice model making use of fuzzy preferences allows actors to make compromises on one dimension for gains in another without necessarily experiencing a loss in utility on the former. In this way, we can predict whether an actor may be more likely to make a compromise for a gain, when the actor perceives no loss in utility on the compromised dimension, while the actor may be more resistant to such a trade-off if some real loss in utility occurs as a result.

4.3.1 A Fuzzy Maximal Set Approach in One-Dimensional Models

Our first model applies the fuzzy maximal set to predict coalition formation in single-dimensional policy space. When applied to the government formation model, an aggregation of the maximal set equation determines viable coalitions. The goal of these definitions is to determine a hierarchy of possible coalitions at a given level of maximality, ultimately selecting the coalition with the highest degree of cooperation between each member in the coalition. This process often results in the weakest member of the coalition deciding the coalition for each policy alternative.

The fuzzy preference relation is inserted into a fuzzy version of the maximal set equation along with a fuzzy characteristic function to infer the degree to which an alternative is considered maximal for player i (Dasgupta and Deb 1991; Georgescu 2007).

Georgescu (2007) proposes the following definition for a fuzzy maximal set.

Definition 4.1 (*Georgescu maximal set*) The fuzzy maxima set $M_G(\sigma_i, \rho_i)(x)$ is given by

$$\sigma_i(x) \wedge \bigwedge_{y \in \text{supp}(\sigma_i)} \bigvee \{t \in [0, 1] \mid \sigma_i(x) \wedge \rho_i(y, x) \wedge t \le \rho_i(x, y)\} \ .$$

Definition 4.1 tells us to what degree x is maximal in the universe X given some characteristic function σ_i and fuzzy preference relation ρ_i. Under certain conditions, $M_G(\sigma_i.\rho_i)(x)$ is a choice function, and we can use it to construct a fuzzy preference relation based on the maximal-ness of each alternative.

Georgescu (2007) lays out these conditions.

Definition 4.2 For $M_G(\sigma_i, \rho_i)(x)$ to be a choice function, ρ_i must be reflexive, transitive and strongly total. The following definition of ρ_i meets these conditions when $c = 1$:

$$\rho_i(x, y) = \begin{cases} [\sigma_i(x) - \sigma_i(y) + c] \wedge 1 & \text{when } \sigma_i(x) \ge \sigma_i(y) \\ 1 - [(\sigma_i(y) - \sigma_i(x) + 1 - c) \wedge 1)] & \text{when } \sigma_i(y) \ge \sigma_i(x) \end{cases}.$$

Assuming $c = 1$, $M_G(\sigma_i, \rho_i)(x)$ becomes a fuzzy choice function. When this occurs, $M_G(\sigma_i, \rho_i)(x)$ indicates the degree to which individual i chooses x given some characteristic function σ_i and fuzzy preference relation ρ_i.

A fuzzy choice function allows us to use revealed preference theory to reconstruct individual preference relations based on derived values of $M_G(\sigma_i.\rho_i)(x)$.

Definition 4.3 The following equations are revealed preference formulas between x and y between y and x, respectively. They use maximal set calculations but restricts the domain to just x and y.

(1) $\overline{R_i}(x, y) = M_G(\sigma_i, \rho_i)(x)|_{\{x,y\}}$
(2) $\overline{R_i}(y, x) = M_G(\sigma_i, \rho_i)(y)|_{\{x,y\}}$

4.3 does not give us a fuzzy aggregation rule quite yet. It applies only to an individual within a collective group. However, we can construct characteristic functions and preference relations for some coalition C and then apply 4.3 to these new group values.

The next definition uses 4.1 to calculate a group characteristic function.

Definition 4.4 A group characteristic function, $\sigma_C(x)$, is defined as $\sigma_C(x) = \bigwedge \sigma_c(x), \forall c \in C$.

It should be noted that when $C = N$, then 4.4 is then a social characteristic function over N, not just for a subset of N. Once we have constructed a social characteristic function, we can use 4.2 to construct a social fuzzy preference relation.

Definition 4.5 A social fuzzy preference relation is defined as

$$\rho_C(x, y) = \begin{cases} [\sigma_C(x) - \sigma_C(y) + c] \wedge 1 & \text{when } \sigma_C(x) \geq \sigma_C(y) \\ 1 - [(\sigma_C(y) - \sigma_C(x) + 1 - c) \wedge 1] & \text{when } \sigma_C(y) \geq \sigma_C(x) \end{cases}.$$

The same properties of reflexivity, transitivity and strong totality also hold in 4.5 when $c = 1$. Thus, we can use $\sigma_C(x)$ and $\rho_C(x, y)$ to construct a social fuzzy preference relation.

Definition 4.6 (*social fuzzy preference relation*) A social fuzzy preference relation, $\overline{R_C}$, is defined as follows:

(1) $\overline{R_C}(x, y) = M_G(\sigma_C, \rho_C)(x)|_{\{x,y\}}$
(2) $\overline{R_C}(x, y) = M_G(\sigma_C, \rho_C)(y)|_{\{x,y\}}$

Finally, when $C = N$, $\overline{R_C}$ becomes a social fuzzy preference relation. The aggregation procedure is demonstrated in the following example.

Example 4.7 Let $N = 1, 2, 3$. Let $X = (0.75, 0.25, 0.25), (0.5, 0, 1), (0.5, 0.5, 0.5)$, where $x \in X$ is written as $(\sigma_1(x), \sigma_2(x), \sigma_3(x))$. Let $a = (0.75, 0.25, 0.25)$, $b = (0.5, 0.25, 1)$ and $c = (0.5, 0.25, 0.5)$. Find $\overline{R_C}(x, y)$ and $\overline{R_C}(x, y)$ for all x, y in X.

$\sigma_C(a) = \bigwedge \{0.75, 0.25, 0.25\} = 0.25$
$\sigma_C(b) = \bigwedge \{0.5, 0.25, 1\} = 0.25$
$\sigma_C(c) = \bigwedge \{0.5, 0.5, 0.5\} = 0.5$
The table represents $\rho_N(x, y)$ when $c = 1$ for all x, y in X.

ρ_N	a	b	c
a	1	1	0.75
b	0.75	1	0.50
c	1	1	1

$$\overline{R_C}(a, b) = M_G(\sigma_C, \rho_C)(a)|_{\{a,b\}}$$
$$= 0.25 \wedge \bigwedge \left\{ \bigvee \{t \in [0, 1] \mid 0.25 \wedge 0.75 \wedge t \leq 1\} \mid y = b \right\}$$
$$= 0.25$$

$$\overline{R_C}(b, a) = M_G(\sigma_C, \rho_C)(b)|_{\{a,b\}}$$
$$= 0 \wedge \bigwedge \left\{ \bigvee \{t \in [0, 1] \mid 0.25 \wedge 1 \wedge t \leq 1\} \mid y = a \right\}$$
$$= 0$$

$$\overline{R_C}(a, c) = M_G(\sigma_C, \rho_C)(a)|_{\{a,c\}}$$
$$= 0.25 \wedge \bigwedge \left\{ \bigvee \{t \in [0, 1] \mid 0.25 \wedge 1 \wedge t \leq 0.75\} \mid y = c \right\}$$
$$= 0.25$$

$$\overline{R_C}(a, c) = M_G(\sigma_C, \rho_C)(c)|_{\{a,c\}}$$
$$= 0.5 \wedge \bigwedge \left\{ \bigvee \{t \in [0, 1] \mid 0.5 \wedge 0.75 \wedge t \leq 1\} \mid y = a \right\}$$
$$= 1$$

$$\overline{R_C}(b, c) = M_G(\sigma_C, \rho_C)(b)|_{\{b,c\}}$$
$$= 0 \wedge \bigwedge \left\{ \bigvee \{t \in [0, 1] \mid 0 \wedge 1 \wedge t \leq 0.75\} \mid y = c \right\}$$
$$= 0$$

$$\overline{R_C}(c, b) = M_G(\sigma_C, \rho_C)(c)|_{\{b,c\}}$$
$$= 0.5 \wedge \bigwedge \left\{ \bigvee \{t \in [0, 1] \mid 0.5 \wedge 0.75 \wedge t \leq 1\} \mid y = b \right\}$$
$$= 1$$

4.3.2 A Fuzzy Pareto Set Approach in One-Dimensional Model

Our second model applies the fuzzy Pareto set to predict coalition formation in single-dimensional policy space. The approach argues that the governments formed will be Pareto efficient. Let $N = \{1, ..., n\}$ denote a finite set of players and X denote a set of alternatives. If R is a binary relation on X, we let P denote the strict preference relation associated with R, i.e., $P = \{(x, y) \in R \mid (y, x) \notin R\}$. Let $\mathscr{R}$ denote the set of all binary relations on X that are reflexive, complete, and transitive. Let $\mathscr{R}^n = \{\rho \mid \rho = (R_1, ..., R_n), R_i \in \mathscr{R}, i = 1, ..., n\}$.

Definition 4.8 Let $\rho \in \mathscr{R}^n$. The Pareto set of ρ, $PS_N(\rho)$, is defined to be

$$PS_N(\rho) = \left\{ x \in X \mid \forall y \in Y, \quad \exists i \in N \ y P_i x \Rightarrow \exists j \in N \ x P_j y \right\}.$$

In words, a policy position at point x on the *rile* dimension is Pareto efficient for a coalition L if there exists some member of L which strictly prefers y to x as well as some other member of L which strictly prefers x to y. These points suggest stability in policy; a move away from them would decrease the utility of some member in the coalition, who would not let such policy pass.

Our model predicts a government will form at the fuzzy Pareto efficient position within the intersection of a legislative coalition's fuzzy preferences. In the model, a legislative coalition is a subset of the entire set of political parties in the parliament whose fuzzy preferences intersect at some policy position and who control at least an absolute majority of the seats. The parties in a legislative coalition will support the formation of a government at the Pareto efficient policy position within the intersection of their preferences. Because the fuzzy model introduces so-called "thick" indifference into parties' preferences (see Sloss 1973; Penn 2006), there is a natural tendency for outcomes to move towards Pareto efficient policy positions (Casey et al. 2012). Our model predicts that coalitions will form at Pareto efficient points. This implies that parties which intersect at higher alpha levels will be more likely to form a coalition than parties which intersect at lower alpha levels. The Pareto efficient coalition then has more room for negotiation on policy as they likely agree on multiple issues.

To determine the Pareto efficiency of a prospective coalition, the intersection of parties' preferences must first be determined. Figure 4.4 depicts an example of two parties whose fuzzy preference profiles intersect at the highest alpha level, 1.00. This example shows the difference in utility gained by each party in the coalition due to a proposed policy shift. At policy point a, both parties receive their highest utility ($\alpha = 1$). However, a movement towards policy point b, would decrease the utility for the party on the right to 0.75. Thus the model would predict that a policy shift would be rejected.

The model also offers simple examples of the types of government that will be predicted to form based on the different kinds of intersections of the fuzzy preference profiles. Figure 4.5 is an example of a two-party minimum winning coalition. In this example, three parties are recognized in the election and each party is assumed to have equal legislative weight. Party A and Party B in this example intersect at $\alpha = 0.75$, but the level at which they intersect is not needed. This is because there are no other cases of intersection present. For this example, the Fuzzy model would predict a coalition to form with Party A and Party B as the members. However, it is also possible to form a government from a subset of the predicted coalition, so minority governments of Party A alone or Party B alone would also be predicted.

Figure 4.6 introduces a case where an intersection of fuzzy preference profiles appears in two places along the policy dimension. In this model, a coalition would be predicted to form with parties A and B, or with parties B and C. Note that the fuzzy preferences of parties A and C do not intersect, so the model would not predict a coalition to form with these parties. If we include the subsets of the predicted

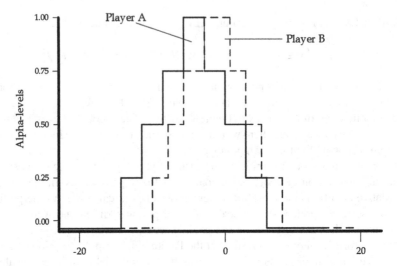

Fig. 4.4 Preference overlap

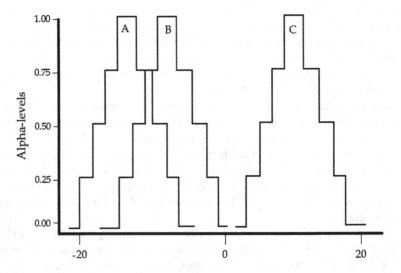

Fig. 4.5 Two-party minimum winning coalition

coalition, minority governments would also be included in the prediction set for each of the three parties.

Figure 4.7 is an example of a three-party supermajority coalition. This coalition may occur when all three parties' fuzzy preferences intersect. The prediction set in this case would be a coalition containing parties A, B, and C. The subset of potential coalitions would be with parties A and B, A and C, and B and C. Single minority governments of each party would also be included in the prediction set.

The previous examples assume equal legislative weight for the parties. Figure 4.8 has the same fuzzy preference intersection as Fig. 4.7, but it assigns different weights

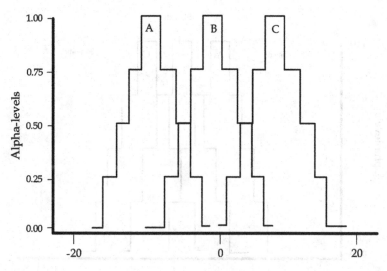

Fig. 4.6 Two possible two-party minimum winning coalitions

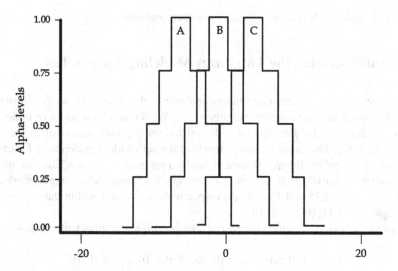

Fig. 4.7 Three-party supermajority coalition (Casey 2009)

to each party in a 100-seat legislature. Party A holds 25 seats, Party B holds 51 seats, and party C holds 24 seats. While Party B can form a government on its own, it may also form a supermajority government with A and C, or even with A or C. The fuzzy preference intersections also add the minority governments of A, C, or A and C to the prediction set.

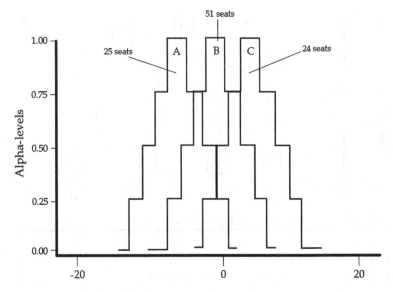

Fig. 4.8 Potential single party majority or supermajority coalitions

4.4 Demonstrating the Two Fuzzy Modeling Approaches

Before we test the two fuzzy one-dimensional public choice models, we demonstrate how they work and the conclusions they reach. We do so with a series of notional situations depicting the intersection of discretized fuzzy preferences of three parties, $N = \{A, B, C\}$. The fuzzy inclusion levels in the set of ideal preferences for each party along the policy dimension are denoted by a series of rectangles. The outermost rectangle represents the 0.50 set level for the respective party. The rectangles marking the limits of the 0.75 and 1 set levels respectively are nested within the outermost rectangle, $T^4 = \{1, 0.75, 0.5, 0\}$.

The set of regions at which the parties' fuzzy preferences intersect in Fig. 4.9 is

$$X = \{(1, 0.5, 0), (1, 0, 0), (0.75, 0.5, 0), (0.75, 0, 0),$$
$$(0.5, 0.5, 0), (0, 1, 0), (0, 0.75, 0), (0, 0.5, 1),$$
$$(0, 0.5, 0.75), (0, 0.5, 0.5), (0, 0.5, 0)(0, 0, 1),$$
$$(0, 0, 0.75), (0, 0, 0.5), (0.5, 0, 0), (0, 0, 0)\}.$$

The fuzzy maximal set is

$$\{(1, 0.5, 0), (0.75, 0.5, 0), (0.5, 0.5, 0),$$
$$(0, 0.5, 1), (0, 0.5, 0.75), (0, 0.5, 0.5),$$
$$(0, 1, 0), (0, 0.75, 0), (0, 0.5, 0)\}$$

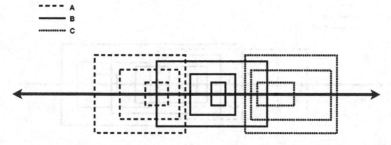

Fig. 4.9 First notional three-party system

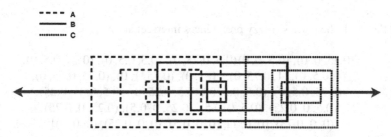

Fig. 4.10 Second notional three-party system

The fuzzy Pareto set is

$$\{(1, 0.5, 0), (0, 1, 0), (0, 0.5, 1)\}$$

The fuzzy maximal set and the fuzzy Pareto set include Player B in every coalition. Thus Player B will not only be in the final solution set, but it will likely determine who will join it in that set.

In Fig. 4.10, the parties' fuzzy preferences intersect at

$$X = \begin{aligned} \{(1, 0.5, 0), (1, 0, 0), (0.75, 0.75, 0), (0.75, 0.5, 0), (0.75, 0, 0), (0.5, 1, 0), \\ (0.5, 0.75, 0), (0, 1, 0), (0, 0.75, 0.5), (0, 0.75, 0), (0, 0.5, 1), (0, 0.5, 0.75), \\ (0, 0.5, 0.5), (0, 0, 1), (0, 0, 0.75), (0, 0, 0.5), (0.5, 0, 0), (0, 0, 0)\} \,. \end{aligned}$$

The fuzzy maximal set is

$$\{(0.75, 0.75, 0), (0.5, 1, 0), (0.5, 0.75, 0), (0, 1, 0)\}$$

The fuzzy Pareto set is

$$\{(1, 0.5, 0), (0.75, 0.75, 0), (0.5, 1, 0), (0, 1, 0), (0, 0.5, 1)(0, 0.75, 0.5)\}$$

Player B is once again likely to decide the outcome, but that outcome will most likely include player A in a coalition.

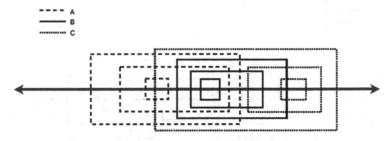

Fig. 4.11 Third notional three-party system

In Fig. 4.11, the parties' fuzzy preferences intersect at

$$X = \{(0.25, 0, 0), (0.5, 0, 0)(0.5, 0.25, 0), (0.5, 0.5, 0), (0.75, 0.5, 0),$$
$$(1, 0.5, 0), (1, 0.75, 0), (1, 1, 0), (0.75, 1, 0), (0.75, 0.75, 0),$$
$$(0.75, 0.5, 0), (0.5, 0.5, 0), (0.5, 0.25, 0), (0.5, 0.25, 0.25),$$
$$(0.5, 0, 0.25), (0.5, 0, 0.5), (0.25, 0, 0.5), (0.25, 0, 0.75),$$
$$(0, 0, 0.75), (0, 0, 1), (0, 0, 0.5), (0, 0, 0.25), (0, 0, 0)\} \ .$$

The fuzzy maximal set is

$$\{(0.75, 1.5), (0.75, 1, 0), (0.75, 0.5, 0.5), (0.75, 0.75, 0.5)\} \ .$$

The fuzzy Pareto set is

$$\{(1, 0, 0.5), (0.75, 1, 0.5), (0, 0.5, 1), (0, 0.75, 0.75)\} \ .$$

In Fig. 4.12, the parties' fuzzy preferences intersect at

$$X = \{(0.5, 0, 0), (0.75, 0, 0), (1, 0, 0), (1, 0.5, 0), (0.75, 0.5, 0),$$
$$(0.75, 0.75, 0), (0.75, 1, 0), (0.5, 1, 0), (0.5, 1, 0.5),$$
$$(0.5, 0.75, 0.5), (0, 0.75, 0.5), (0, 0.75, 0.75), (0, 0.75, 1),$$
$$(0, 0.5, 1), (0, 0, 1), (0, 0, 0.75), (0, 0, 0.5)\}$$

The fuzzy maximal set is

$$\{(0.75, 1, 0), (0.5, 1, 0), (0.5, 1, 0.5), (0.5, 1, 0)\} \ .$$

The fuzzy Pareto set is

$$\{(1, 0.5, 0), (0.75, 1, 0), (0.5, 1, 0.5), (0, 0.75, 0.1)\} \ .$$

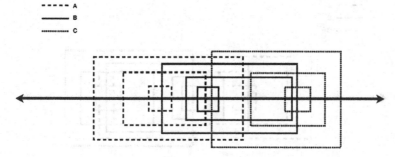

Fig. 4.12 Fourth notional three-party system

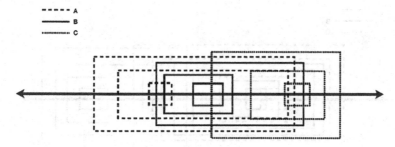

Fig. 4.13 Fifth notional three-party system

In Fig. 4.13, the parties' fuzzy preferences intersect at

$$X = \begin{Bmatrix} (0.5, 0, 0), (0.75, 0, 0), (1, 0, 0), (1, 0.5, 0), (1, 0.75, 0), \\ (0.75, 0.75, 0), (0.75, 1, 0), (0.75, 1, 0.5,), (0.75, 0.75, 0.5), \\ (0.75, 0.5, 0.5), (0.75, 0.5, 0.75), (0.75, 0.5, 1), (0.5, 0.5, 1), \\ (0, 0.5, 1), (0, 0, 1), (0, 0, 0.75), (0, 0, 0.5) \end{Bmatrix}.$$

The fuzzy maximal set is

$$\{(0.75, 10), (0.75, 1, 0.5), (0.75, 0.75, 0.5)\}.$$

The fuzzy Pareto set is

$$\{(1, 0.75, 0), (0.75, 1, 0.5), (0.75, 0.5, 1)\}.$$

In Fig. 4.14, the parties' fuzzy preferences intersect at

$$X = \begin{Bmatrix} (0.5, 0, 0), (0.75, 0, 0), (1, 0, 0), (1, 0.5, 0), \\ (0.75, 0.5, 0)(0.75, 0.75, 0), (0.75, 1, 0), (0.75, 1, 0.5,), \\ (0.75, 0.75, 0.5), (0.75, 0.5, 0.5), (0.75, 0.5, 0.75), (0.75, 0.5, 1), \\ (0.5, 0.5, 1), (0.5, 0, 1), (0, 0, 1), (0, 0, 0.75), (0, 0, 0.5) \end{Bmatrix}.$$

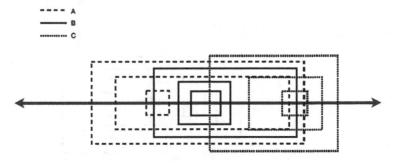

Fig. 4.14 Sixth notional three-party system

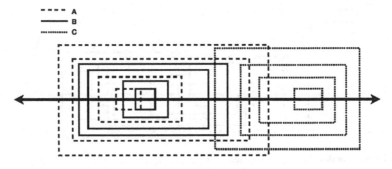

Fig. 4.15 Seventh notional three-party system

The fuzzy maximal set is

$$\{(0.75, 1, 0.5), (0.75, 0.75, 0.5), (0.75, 0.5, 0.5), (0.75, 0.5, 0.75), (0.75, 0.5, 1)\}.$$

The fuzzy Pareto set is

$$\{(1, 0.5, 0), (0.75, 1, 0.5), (0.75, 0.5, 1)\}.$$

In Fig. 4.15, the parties' fuzzy preferences intersect at

$$\{(0.25, 0, 0), (0.5, 0, 0)(0.5, 0.25, 0), (0.5, 0.5, 0), (0.75, 0.5, 0),$$
$$(1, 0.5, 0), (1, 0.75, 0), (1, 1, 0), (0.75, 1, 0), (0.75, 0.75, 0),$$
$$X = (0.75, 0.5, 0), (0.5, 0.5, 0), (0.5, 0.25, 0), (0.5, 0.25, 0.25), (0.5, 0, 0.25), .$$
$$(0.5, 0, 0.5), (0.25, 0, 0.5), (0.25, 0, 0.75), (0, 0, 0.75),$$
$$(0, 0, 1), (0, 0, 0.5), (0, 0, 0.25), (0, 0, 0)\}$$

The fuzzy maximal set is

$$\{(1, 1, 0), (1, 0.5, 0), (1, 0.75, 0), (0.75, 1, 0)\}.$$

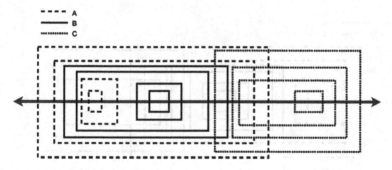

Fig. 4.16 Eighth notional three-party system

The fuzzy Pareto set is

$$\{(1, 1, 0), (0.5, 0.25, 0.25), (0.5, 0, 0.5), (0.25, 0, 0.75), (0, 0, 1)\}.$$

In this scenario, Player A will get its ideal point at $\{1, 1, 0\}$.
   In Fig. 4.16, the parties' fuzzy preferences intersect at

$$
\begin{aligned}
& \{(0.25, 0, 0), (0.5, 0, 0)(0.5, 0.25, 0), (0.5, 0.5, 0), \\
& \quad (0.75, 0.5, 0), (1, 0.5, 0), (0.5, 0.75, 0), (0.5, 1, 0), \\
X = {} & (0.5, 0.25, 0)(0.5, 0.25, 0.25), (0.5, 0, 0.25), (0.5, 0, 0.5), . \\
& \quad (0.5, 0, 0.75)(0.25, 0, 0.75), (0, 0, 0.75), \\
& \quad (0, 0, 1), (0, 0, 0.5), (0, 0, 0.25), (0, 0, 0)\}
\end{aligned}
$$

The fuzzy maximal set is

$$\{(1, 0.5, 0), (0.75, 0.5, 0), (0.5, 0.5, 0), (0.5, 1, 0), (0.5, 0.75, 0)\}.$$

The fuzzy Pareto set is

$$\{(1, 0.5, 0), (0.5, 1, 0), (0.5, 0.25, 0.25), (0.5, 0, 0.75), (0, 0, 1)\}.$$

In this scenario, it is likely that either Player A or Player B will achieve their best
outcome at set inclusion level of 1.
   In Fig. 4.17, the parties' fuzzy preferences intersect at

$$
\begin{aligned}
& \{(0.25, 0, 0), (0.5, 0, 0), (75, 0, 0), (1, 0, 0), \\
& \quad (0.5, 0.25, 0), (0.5, 0.5, 0), (0.5, 0.75, 0), (0.5, 1, 0), \\
X = {} & (0.5, 0.25, 0.25), (0.5, 0, 0.25), (0.5, 0, 0.5), (0.5, 0, 0.75), . \\
& \quad (0.25, 0, 0.75), (0, 0, 0.75), (0, 0, 1), \\
& \quad (0, 0, 0.5), (0, 0, 0.25), (0, 0, 0)\}
\end{aligned}
$$

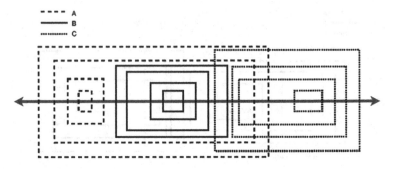

Fig. 4.17 Ninth notional three-party system

The fuzzy maximal set is

$\{(0.5, 1, 0), (0.5, 0.75, 0), (0.5, 0.5, 0), (0.5, 0.25, 0), (0.5, 0.25, 0.25), (0.5, 0, 0.75),$
$\quad (0.5, 0, 0.5), (0.5, 0, 0.25)\}$.

The fuzzy Pareto set is

$$\{(1, 0, 0), (0.5, 1, 0), (0.5, 0.25, 0.25), (0.5, 0, 0.75), (0, 0, 1)\}.$$

While in this scenario, Player B is the only one able to achieve an outcome at its highest set inclusion value, Player A will decide the outcome.

In Fig. 4.18, the parties' fuzzy preferences intersect at

$$X = \begin{matrix} \{(0.25, 0, 0), (0.5, 0, 0), (75, 0, 0), (1, 0, 0), (0.5, 0.25, 0), \\ (0.5, 0.5, 0), (0.5, 0.75, 0), (0.5, 1, 0), (0.5, 0.25, 0.25), (0.5, 0, 0.25), \\ (0.5, 0, 0.5), (0.5, 0, 0.75), (0.5, 0, 1), (0.25, 0, 1), \\ (0, 0, 0.75), (0, 0, 1), (0, 0, 0.5), (0, 0, 0.25), (0, 0, 0)\} \end{matrix}.$$

The fuzzy maximal set is

$$\{(0.5, 1, 0), (0.5, 0.75, 0), (0.5, 0.5, 0), (0.5, 0.25, 0),$$
$$(0.5, 0.25, 0.25), (0.5, 0, 1), (0.5, 0, 0.75), (0.5, 0, 0.5), (0.5, 0, 0.25)\}.$$

The fuzzy Pareto set is

$$\{(1, 0, 0), (0.5, 1, 0), (0.5, 0.25, 0.25), (0.5, 0, 0.1)\}.$$

In this scenario, Players B and C might achieve an outcome at their highest set inclusion value. However, Player A will decide the outcome.

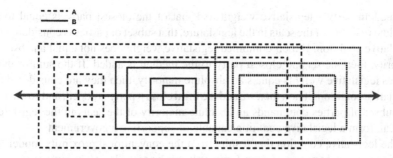

Fig. 4.18 Tenth notional three-party system

4.5 Comparing the Fuzzy and Conventional Models of Government Formation

To test the performance of our fuzzy public choice models, we compare their predictions against those of two conventional models. The first of the conventional models makes use of the Median Voter Theorem. The basic principle is that parties on either side of an issue dimension attempt to draw policy as closely to their ideal positions as possible. Since some quorum, usually a majority, must be achieved, policy gets pulled to the position of the median voter, where a veritable tug-of-war of policy preferences holds policy.

The median voter model identifies likely governing coalitions by identifying legislative coalitions on either end of the issue dimension. Starting with the left-most party on the issue dimension, the median moves from left to right, adding each party's legislative weight, until the sum of the parties' legislative weights is enough to achieve an absolute majority. The right-most party in this left-wing subset of parties is the median voter from the left. Then the same is done from the right-most party leftward. The left-most party in this subset of right-wing parties is the median voter from the right. More often than not, the median voter model identifies a single-party as the median voter from both wings. However, there are occasions when the median voter model identifies two parties as the median. In such a case, we assume that one or both of the parties will be the government. For this reason, the median voter model is most likely to predict single-party majority or minority governments, and only occasionally predicts two-party coalition governments. The model is not able to predict a supermajority coalition.

Because the median voter model is limited in its ability to predict government formation, it would be uncharitable to compare it to the fuzzy model as the sole application of the conventional public choice model. Therefore, we compare the fuzzy model to another simple application of the Euclidean model called the *crisp proximity model*. The crisp proximity model is an extension of the median voter model. First, the median voter is identified. If the median voter is not a single party, the median point on the *rile* dimension between the two median parties is treated as the median. Then, the closest party to the median is identified. If the sum of the median party's legislative weight and that of the closest party is equal to an absolute majority of the seats in the legislature, that subset of parties is considered the legislative coalition. If their combined legislative weight does not equal an absolute majority, then the second closest party to the median is added. If the sum of these parties legislative weights equals an absolute majority, then they are considered the legislative coalition. If not, then we add the third closest party is added, and so on until the subset of parties commands an absolute majority of the seats in the legislature and can form a legislative coalition capable of supporting a government.

The legislative coalition that results from the conventional proximity model can be of a range of shapes and forms. It must always be at the least a minimum-winning coalition. However, the legislative coalition predicted by this model can also be a supermajority coalition. If the final party added to the coalition allows it to achieve an absolute majority and also pushes its legislative weight to the point that some other party can be removed without losing the majority, the coalition remains intact and is treated as a supermajority coalition.

Once the legislative coalition is identified, any subset of the parties within the coalition is considered a possible governing coalition. If the legislative coalition is a minimum-winning coalition of parties, then either party is treated as a potential single-party minority government and the two together are treated as a potential minimum-winning coalition government. If the legislative coalition is a minimum-winning coalition of three parties, then each party is a potential single-party minority, any combination of two can form a minority coalition, or the three together can be a minimum-winning coalition government. If the resulting legislative coalition is a supermajority of three parties, A, B, and C, such that a can form a minimum-winning coalition with either B or C, but B and C can only achieve a minority with each other, then three single-party minority governments are possible, along with a minority coalition, two minimum-winning coalitions, and a supermajority coalition of all three.

The crisp proximity model is clearly a very flexible application of the conventional public choice model to group decision-making. Like our simple fuzzy model, the crisp proximity model identifies a legislative coalition from which any subset of parties can form a government. However, because its predictions rely on the median voter model, the crisp proximity model can only predict a single legislative coalition. Furthermore, if a single party commands an absolute majority of the legislative seats, the crisp proximity model can only predict a single-party majority government. In contrast, our fuzzy models can predict that a single party with an absolute majority can form supermajority coalitions with other parties if the parties' fuzzy preference profiles intersect. While such behavior is rare in government formation, it is possible and not unheard of (e.g., Germany in 1957).

4.6 Results

Table 4.1 reports the predictions made by the four models of 235 real-world governments. Of the governments included in the analysis, 44 (18.7 %) are single-party minority governments, 27 (11.5 %) are minority coalitions, 23 (10.2 %) are single-party majority governments, 89 (37.5 %) were minimum-winning coalitions, and 52 (22.1 %) are supermajority governments.

Overall, the fuzzy Pareto set model performed substantially better than the other three models. The fuzzy Pareto set model predicted 129 (54.9 %) governments correctly, the conventional proximity model predicted 98 governments (41.7 %) governments correctly, the fuzzy maximal set model predicted 87 (37.0 %) of the governments correctly, and the conventional median voter model predicted 46 (19.6 %) of the governments correctly.

Unlike the conventional models, the fuzzy models may fail to make a prediction. This occurs when the intersection of parties' preferences do not include a set of parties with a legislative majority in the assembly. This occurred in 39 instances (16.6 % of the governments), and it affected both fuzzy models equally as they both rely on the same fuzzy preferences. If these cases are taken out of the calculations, 43.8 % of the predictions made by the fuzzy maximal set model were correct and 65.8 % of the predictions made by the fuzzy Pareto set model are correct.

While the performance of the fuzzy maximal set model is a bit disappointing in comparison with the conventional proximity model, we note that the fuzzy maximal set model generates more than one prediction. Table 4.1 reports a prediction as correct only if it is the coalition with the highest set inclusion in the maximal set. When we include governments that are predicted at lower set inclusion levels in the results, the fuzzy maximal set model predicts 120 (51.1 %) of the governments correctly, which includes 23 (100 %) of the single-party majority governments, 31 (69.0 %) of the single-party minority governments, 12 (44.4 %) of the minority coalitions, 38 (42.7 %) of the minimum-winning coalitions, and 16 (30.7 %) of the supermajority governments. This surpasses the prediction rate of the conventional proximity model in all one category, single-party minority governments.

In general, we are quite satisfied with the performance of our fuzzy models, particularly that of the fuzzy Pareto set model. Our satisfaction is increased by the realization that the preference horizon (support) of the parties' fuzzy preferences may have in some cases affected the prediction rate of the fuzzy models adversely. In the process of re-shaping the density functions resulting from the bootstrap, the support of parties' fuzzy preference profiles are foreshortened to create rectangular α-levels. This shortening of fuzzy preference profiles may eliminate intersections that would exist otherwise. Increasing the range of the support of parties' fuzzy preference profiles may identify other potential legislative coalitions.

Table 4.1 Summary of results

	Conventional median voter model	Conventional proximity model	Fuzzy maximal set model	Fuzzy pareto set model
Single-party minority	23 (52.3%)	35 (79.6%)	30 (66.7%)	34 (77.3%)
Minority coalition	0 (0.0%)	8 (29.6%)	7 (25.9%)	14 (51.9%)
Single-party majority	23 (100%)	23 (100%)	23 (100%)	23 (100%)
Minimum winning coalition	0 (0.0%)	26 (29.2%)	21 (20.2%)	39 (43.8%)
Supermajority	0 (0.0%)	6 (11.5%)	6 (11.5%)	19 (36.5%)
Total	46 (19.6%)	98 (41.7%)	87 (37.0%)	129 (54.9%)

4.7 Conclusions

The results indicate that the fuzzy public choice model is a viable alternative to the conventional model. Not only does the fuzzy model release us from many of the assumptions of the conventional model and allow us to make stable predictions of decision-making outcomes in as single-dimensional space, it also outperforms conventional models in the prediction of government formation. Furthermore, the fuzzy model allows us to utilize important sources of preference data such as the CMP despite the prevalence of unbiased error in the data. In fact, the fuzzy model utilizes such error to determine the position, size, and shape of parties' preferences and whether parties agree at the intersections of their preferences. On the other hand, unbiased error found by BLM using their bootstrap procedure makes CMP data, and perhaps also other text data, almost useless for the sake of identifying political actors' policy positions.

Our results also argue that fuzzy public choice models present a fresh approach to the study of collective decision-making. Moreover, the fuzzy preference measures developed in this study are an exciting first step towards their further use. While conventional public choice models still dominate the literature, the overly precise measurement of players' ideal points renders many of these models clumsy.

Indeed, many of the advantages of the fuzzy public choice model owe to the fact that the model is released from the restrictions of the Euclidean distance assumption. A fuzzy model allows us to represent players' preferences as subsets of policy space over which players do not perceive minute shifts in utility with every incremental shift in policy. Rather players may have large areas of the policy space over which they are indifferent to policy shifts and perceive shifts in utility at different granularities from other players. That the fuzzy model treats players' preferences as subsets of policy over which they perceive utility shifts rather than distances between ideal points allows the fuzzy model to make a more diverse set of predictions than the conventional Euclidean distance approach. While the decision-making predictions of the Euclidean model are bound to the median policy position, the fuzzy model's predictions may include agreements made towards both the median and the extremes of the dimensions, depending on the intersection of players' preference profiles. This is a distinct advantage of the fuzzy model, and why it may be a viable alternative to the traditional public choice model.

A disadvantage of the application of the fuzzy model to government formation in this study could be its lack of precision. While the fuzzy model's advantage is that it can make a diverse set of predictions about potential legislative, and therefore governing, coalitions, it also serves as a disadvantage as the wide range of predicted coalitions in some instances may make the model's predictions somewhat arbitrary. However, more often than not, the fuzzy models fall far short of predicting the entirety of potential coalitions, at least narrowing down the set of possible coalitions.

Inherent in the fuzzy model is the possibility of making more specific predictions. For example, the α-levels at which parties' preference profiles intersect could be used to determine the likelihood that they will form a coalition. If we assume parties are unitary, rational and policy-motivated, then a party with the option to form two coalitions, one in which it receives a utility $\alpha = 1.00$ and the other in which it receives a utility $\alpha = 0.75$, is more likely to agree to the first. Adding such considerations to the application of the fuzzy model to government formation could either narrow the set of coalitions we consider viable or provide a way of calculating the likelihood of their formation. The latter approach is a substantial improvement over the conventional model, which either predicts that the coalition forms or does not form.

Our fuzzy models also raise some questions about our use of discrete rather than continuous fuzzy preferences. When determining the formation of coalitions, we did not make full use of the α-levels at which parties' preferences intersect. Rather, we simply took an intersection of parties' preferences as an indication of a potential coalition, and did not take utility into account. This was particularly true for the fuzzy Pareto set model. In a sense, our study was binary; zero versus non-zero utility, with an assumption that parties would form form a coalition if their utility was simply non-zero.

Despite the success of the fuzzy model, we are left with the challenge of having failed to predict a goodly number of governments. Governments not predicted correctly by the fuzzy model could be accounted for by error within our fuzzy preference measures, within the bootstrap procedure, within the CMP data, within our application of the fuzzy model, within the fuzzy model itself, or it could simply account for the rate at which parties' preferences are an accurate predictor of political behavior. Because this is unclear, it is difficult to deduce the accuracy of our preference measures. In order to deduce their accuracy, an appropriate loss function must be specified. At this point, loss functions have not been explored to any great extent by scholars for either fuzzy or Euclidean public choice models. More often than not, predictions are simply considered "correct" if correctly predicted by the model or "incorrect" if not predicted, and that is the method used here. This does not account for many "close" predictions or predictions that are predicted correctly by accident despite a misrepresentation of the measure. To improve our understanding of the accuracy of preference measures, we must determine better ways to test their validity.

Thus, while the results in this chapter gives us optimism about the utility of fuzzy public choice models, they also open the door to considering further refinements. Among the most obvious is to extend the fuzzy approach developed thus far in this book to multidimensional public choice models. We turn to this task in the next two chapters.

References

Benoit, K., Laver, M., Mikhalov, S.: Treating words as data with error: estimating uncertainty in the comparative manifesto measures. Am. J. Polit. Sci. **53**(2), 49–513 (2009)

Casey, P.C.: Extracting fuzzy preference measures to predict government formation, MA thesis, Creighton University (2009)

Casey, P., Wierman, M.J., Gibilisco, M.B., Mordeson, J.N., Clark, T.D.: Assessing policy stability in iraq: a fuzzy approach to modeling preferences. Public Choice **151**(3–4), 402–423 (2012)

Ching, S., Serizawa, S.: A maximal domain for the existence of strategy-proof rules. J. Econ. Theor. **78**(1), 157–166 (1998)

Clark, T.D., Larson, J., Mordeson, J.N., Potter, J., Wierman, M.J.: Applying Fuzzy Mathematics to Formal Model in Comparative Politics. Springer, Berlin (2008)

Dasgupta, M., Deb, R.: Fuzzy choice functions. Soc. Choice Welfare **8**, 171–182 (1991)

Georgescu, I.: The similarity of fuzzy choice functions. Fuzzy Sets Syst. **158**, 1314–1326 (2007)

Laver, M., Shepsle, K.: Coalitions and cabinet goverment. Am. Polit. Sci. Rev. **84**(3), 879–890 (1990)

Laver, M., Shepsle, K.: Making and Breaking Government: Cabinets and Legislatures in Parliamentary Democracies. Cambrige University Press, New York (1996)

Martin, L.W., Vanberg, G.: Policing the bargain: coalition government and parlamentary scrutiny. Am. J. Polit. Sci. **48**(1), 13–27 (2004)

Masso, J., Neme, A.: A maximal domain of preferences for the division problem. Game. Econ. Behav. **37**(2), 367–387 (2001)

Müller, W.C., Strøm, K.: Coaltion Governments in Western Europe. Oxford University Press, Oxford (2000)

Penn, E.M.: Alternative definitions of the uncovered set, and their implications. Soc. Choice Welfare **27**(1), 83–87 (2006)

Shugart, M.S., Carey, J.M.: Presidents and Assemblies: Constitutional Design and Electoral Dynamics. Cambridge University Press, Cambridge (1992)

Sloss, J.: Stable outcomes in majorty rule voting games. Public Choice **15**(1), 19–48 (1973)

Theis, M.F.: Keeping tabs on partners: the logic of delegation in coalition governments. Am. J. Polit. Sci. **45**(3), 580–598 (2001)

Tsebelis, G.: Veto Players: How Political Institutions Work. Princeton University Press, Princeton (2002)

Warwick, P.V.: When far apart becomes too far apart: evidence for a threshold effect in coaltion formation. Brit. J. Polit. Sci. **35**(3), 323–401 (2005a)

Warwick, P.V.: Do policy horizons structure the formation of parliamentary governments?: the evidence from an expert survey. Am. J. Polit. Sci. **49**(2), 373–387 (2005b)

Chapter 5
Issues in Fuzzy Multi-dimensional Public Choice Models

Abstract Conventional multi-dimensional public choice models are notoriously unstable. Under all but the most restrictive assumptions, they fail to produce a maximal set under majority rule. We demonstrate that fuzzy multi-dimensional public choice models offer a wider degree of stability without resort to highly complex mathematical calculations.

5.1 Stability in Multi-dimensional Models

Scholars wishing to apply conventional assumptions to multi-dimensional public choice models are confronted with McKelvey's Chaos Theorem. Conventional multi-dimensional public choice models relying on social preference to make predictions are inherently unstable (McKelvey 1976, 1979; Plott 1967; Riker 1980). McKelvey (1976), and later McKelvey and Schofield (1986), show that absent Plott's radial symmetry condition (Plott 1967), any point in space can be chosen through a process of majority cycling, even if it disadvantages every player involved. Thus, multi-dimensional public choices are not able to predict an outcome. This occurs for two reasons. First, collective preferences are highly likely to suffer from intransitivity in multi-dimensional policy settings. That is, there is no set of alternatives that is at least as good as any other alternative. Every alternative can be defeated by a majority by at least one other alternative. Second, actors are myopic. They do not consider the future consequences of their actions.

Scholars have attempted to induce a maximal set in two different ways. One approach has been to reconsider McKelvey's basic assumptions. Enelow and Hinich (1984) show that when actors are risk adverse and the voting rule incorporates some amendment agenda, a prediction set exists. Tovey (1991) introduces an epsilon core, or an area of indifference around actors' winset. He shows that when the N size is sufficiently high, equilibrium exists. A second effort to generate a maximal set has been to consider the ability of institutions to induce a stable outcome. Shepsle

P. C. Casey et al., *Fuzzy Social Choice Models*, Studies in Fuzziness
and Soft Computing 318, DOI: 10.1007/978-3-319-08248-6_5,
© Springer International Publishing Switzerland 2014

(1979) proposes the idea that institutions induce equilibria when they restrict voting to a single dimension or reduce the number of possible alternatives. Schofield (1986) considers a non-collegial voting rule and shows if dimensions are limited, then a core exists. Strom, Budge and Laver (1994) analyze a list of institutional variations across parliamentary democracies and suggest how they affect coalition bargaining. Diermeier et al. (2003) construct a game theoretic model that includes structural features such as positive parliamentarism, the vote of no-confidence and the fixed inter-election period. Laver and Shepsle (1996) assume that cabinet ministers dictate coalition policy corresponding to their jurisdiction. (There is increasing research suggesting this is not always the case Warwick 1999; Theis 2001.)

Besides efforts to generate a maximal set, scholars have also attempted to limit alternatives to subregions of the multi-dimensional policy space in public choice models. One of the most interesting approaches in this vein is offered by the uncovered set Miller (2007); Penn (2006). The approach essentially argues that under the assumption that players are strategic and that their choices are reached by way of an amendment agenda, the outcome will be restricted to a proper subset of the Pareto set (Miller 1980; Shepsle and Weingast 1987; McKelvey 1986). Noting that when players' preferences include indifference, differing definitions of the uncovered set return differing results in the conventional model, Mordeson et al. (2011) develop a definition of a fuzzy covering relation, which in most cases results in outcomes being confined to the Pareto set.

The inclusion of indifference in actors' preferences represents a relatively underdeveloped approach to the problem of intransitivity in public choice models. Game-theoretic and social choice studies have shown that models with indifference have a higher probability of producing a nonempty maximal set (Sloss 1973; Tovey 1991; Gehrlein and Valognes 2001; Ehlers and Barbera 2007). Three important conclusions emerge from this literature. First, when formal models allow actors to be indifferent between some alternatives, the probability of an alternative being undefeated increases (Gehrlein and Valognes 2001; Ehlers and Barbera 2007). Second, as the number of players who exhibit indifference over alternatives (even at fine granularities) increases to infinity, the probability of an undefeated alternative existing approaches one (Tovey 1991). Third, relatively small degrees of indifference greatly increase the probability that social choice is stable and centrally located when the number of actors is finite (Koehler 2001).

One approach to incorporating indifference in public choice models is the epsilon core (Tovey 1991; Koehler 2001; Brauninger 2007). The epsilon core is some value, $\varepsilon > 0$, such that two alternatives must differ by ε distance for actors to distinguish between the two. If two alternatives are closer than some ε distance, players are indifferent between the two. Disappointingly, however, epsilon cores require fairly high N-sizes to produce a stable outcome. While Koehler (2001) concludes that "stability can be achieved at relatively low individual preference thresholds" (p. 166), only committees larger than 21 members are tested. Preference thresholds must subsequently be increased to induce stability with less actors. Considering government formation usually includes a smaller N-size, the ε would have to be increased substantially to ensure a stable outcome in models with only three or four actors (Brauninger 2007).

Another approach to incorporating indifference in public choice models is to assume actors have discretized utility functions. In other words players possess "thick" indifference. Sloss (1973) argues decision costs are common to voting and to forming coalitions to change the status quo. As a result, an actor will not exert the effort to choose an alternative over another if the two are separated by a marginal difference in utility. While Sloss demonstrates that these models produce a stable outcome, her models were never applied empirically. One reason for this is that her approach requires the use of differentiable utility functions, which are then used to derive directional gradients to determine stable points. Differential equations are difficult to create and manipulate, making Sloss's argument difficult to empirically test. The ultimate complexity of the model outweighs its usefulness and does not lend itself to empirical verification.

Mordeson and Clark (2010, 2012) presents a fuzzy approach to modeling thick indifference that permits relative ease in calculating prediction sets. Following the approach used in this book, they use discrete fuzzy numbers to represent the preferences of political actors. Their main theorem demonstrates that in all but a limited number of cases, multi-dimensional fuzzy public choice models result in an empty maximal set if and only if the Pareto set contains a union of cycles. Otherwise, these same models result in a nonempty maximal set if the Pareto set does not contain a cycle. Moreover, the fuzzy set approach to modeling thick indifference in Mordeson and Clark (2010) can accommodate highly irregular-shaped indifference curves, even those that are concave or multi-modal. Its ability to do so owes to a homomorphism that permits a region of interest to be mapped to a simpler region with a suitable and natural partial ordering where the results are determined and then faithfully transferred back to the original region of interest.

This chapter will consider the effect of introducing thick indifference in public choice models by using discretized fuzzy preferences. We repeat the work in Mordeson and Clark (2010, 2012) here in order to provide complete enough treatment of the issues involved in applying fuzzy logic to multi-dimensional public choice models. Our intent is to release the reader from having to refer to sources other than this book in order to understand what follows in the next chapter. We conclude with a discussion of fuzzy uncovered relations. While we do not make use of the uncovered solution set in this book, we include the discussion for those readers who might be interested in doing so.

5.2 Majority Rule Maximal Sets

The approach taken in the following section will be quite constructive in nature. In this section we take an axiomatic approach to the same issue. The approach rests on six conditions concerning the partial order. The main result in the current section is similar to Theorems 5.24 and 5.25, but differs in some minor details. The approach in this section also lends itself easily to the study of the uncovered set that follows in the ensuing section.

Here (and later in Sect. 5.4) we will consider conditions under which a majority rule maximal set exists. Our major result is that in all but a limited number of cases, multi-dimensional public choice models of individual preferences of thick indifference result in an empty majority rule maximal set if and only if the Pareto set contains a union of cycles. Moreover, Theorem 5.15, makes it clear that it is the intersections of the players' preferences that matters, rather than the shape of their preferences in multi-dimensional spatial representation.

Let N denote the set of players and X denote the set of alternatives. We assume that X is a subset of a universe U of interest. Recall that $\mathscr{R}$ is the set of all binary relations on X which are reflexive, complete and transitive. Let $\mathscr{R}^n = \{\rho \mid \rho = (R_1, \ldots, R_n), R_i \in \mathscr{R}, i = 1, \ldots, n\}$, where $|N| = n$.

Property 5.1 *Let $\preceq$ be a partial order on U. Let $\rho \in \mathscr{R}^n$. Suppose that $\preceq$ satisfies the following properties:*

(1) $\forall x, y \in U, x \preceq y$ implies $\forall i \in N, yR_i x$;
(2) $\forall x, y, z \in U, \forall i \in N, x \preceq y$ and $xR_i z$ implies $yR_i z$;
(3) $\forall x, y, z \in U, \forall i \in N, x \preceq y$ and $xP_i z$ implies $yP_i z$;
(4) $\forall x, y \in U, x \prec y$ implies $\exists i \in N$ such that $yP_i x$;
(5) $\forall x, y, z \in U, \forall i \in N, x \preceq y$ and $zR_i y$ implies $zR_i x$;
(6) $\forall x, y \in U, x$ and y incomparable under $\preceq$ implies $\exists i \in N$ such that $xP_i y$ implies $\exists j \in N$ such that $yP_j x$.

Definition 5.2 Let $\rho \in \mathscr{R}^n$. Define the binary relation R on X by $\forall x, y \in X, (x, y) \in R$ if and only if $|\{i \in N \mid xR_i y\}| \geq n/2$. Define $P \subseteq X \times X$ by $\forall x, y \in X, (x, y) \in P$ if and only if $(x, y) \in R$ and $(y, x) \notin R$. Let $R(x, y; \rho) = \{i \in N \mid xR_i y\}$ and $P(x, y; \rho) = \{i \in N \mid xP_i y\}$.

In the following R is defined as in Definition 5.2.

Proposition 5.3 *Let $x, y \in X$. Then $(x, y) \in P$ if and only if $|P(x, y; \rho)| > n/2$.*

Proof $xPy \Leftrightarrow xRy$ and not $yRx \Leftrightarrow |\{i \in N \mid xR_i y\}| \geq n/2$ and $|\{j \in N \mid yR_j x\}| < n/2$. Since each R_i is complete, R is complete. Hence $xPy \Leftrightarrow |\{i \in N \mid xP_i y\}| > n/2$ by a simple counting procedure. Thus, $xPy \Leftrightarrow |P(x, y; \rho)| > n/2$.

Definition 5.4 Let $\rho \in \mathscr{R}^n$. The Pareto set of R with respect to ρ is the set

$$PS_N(R) = \left\{x \in X \mid \forall y \in X \left(\exists i \in N \; yP_i x \Rightarrow \exists j \in N \; xP_j y\right)\right\}.$$

Definition 5.5 The maximal set of R with respect to X is the set

$$M(R, X) = \{x \in X \mid \forall y \in X, xRy\}.$$

Recall that $\forall x, y \in U, x \prec y$ if and only if $x \preceq y$ and $x \neq y$.

Definition 5.6 (*Undominated*) We define the set M_R to be those elements that are undominated with respect to the partial order $\prec$.

$$M_R = \{x \in X \mid \nexists y \in X, \ x \prec y\}.$$

Proposition 5.7 $M_R = PS_N(R)$.

Proof Suppose $x \in M_R$. Let $y \in X$. Suppose $\exists i \in N$ such that yP_ix. Now there does not exist $y \in X$ such that $x \prec y$. Thus $\forall y \in X$, either $y \preceq x$ or x and y are not comparable. Since yP_ix, $y \preceq x$ is impossible else $xR_iy\forall i \in N$ by (P1). Hence x and y are incomparable under $\preceq$. Thus $\exists j \in N$ such that xP_jy by ((6)). Hence $x \in PS_N(R)$. Thus $M_R \subseteq PS_N(R)$.

Suppose $x \in PS_N(R)$. Suppose there exists $y \in X$ such that $x \prec y$. Then $\exists i \in N$ such that yP_ix. Since $x \in PS_N(R)$, there exists $j \in N$ such that xP_jy. Thus $x \prec y$ is impossible. Hence $x \in M_R$. Therefore $PS_N(R) \subseteq M_R$.

Corollary 5.8 *Let* $x \in X$.

(i) *Suppose* $\forall y \in X, x \preceq y$ *implies* $x = y$. *Then* $x \in PS_N(R)$.
(ii) *If* $x \notin PS_N(R)$, *then there exists* $y \in PS_N(R)$ *such that* $x \prec y$.

Proof (i) Clearly $x \in M_R$, but $M_R = PS_N(R)$.
(ii) Since $x \notin PS_N(R)$, $x \notin M_R$. Thus there exists $y \in X$ such that $x \prec y$. Let y be the largest such element. Then $y \in M_R = RS_N(R)$.

Definition 5.9 Define $\langle\rangle : \mathscr{P}(U) \to \mathscr{P}(U)$ by $\forall S \in \mathscr{P}(U)$, $\langle S \rangle = \{x \in U \mid \exists s \in S, x \preceq s\}$.

We note that the next result has been used in the study of automata theory and graph theory (Mordeson and Nair 1996).

Proposition 5.10 *Let* $\langle\rangle : \mathscr{P}(U) \to \mathscr{P}(U)$ *be defined as above. Then the following conditions hold.*

(i) $\forall S \in \mathscr{P}(U), S \subseteq \langle S \rangle$;
(ii) $\forall S_1, S_2 \in \mathscr{P}(U), S_1 \subseteq S_2$ *implies* $\langle S_1 \rangle \subseteq \langle S_2 \rangle$;
(iii) $\forall S \in \mathscr{P}(U), \langle S \rangle = \langle\langle S \rangle\rangle$;
(iv) $\forall S \in \mathscr{P}(U), \langle S \rangle = \cup_{s \in S} \langle\{s\}\rangle$;
(v) $\forall S \in \mathscr{P}(U), \forall x, y \in X, x \in \langle S \cup \{y\}\rangle$ *and* $x \notin \langle S \rangle$ *implies* $x \in \langle\{y\}\rangle$.

Proof (i) Let $s \in S$. Then $s \preceq s$ and so $s \in \langle S \rangle$. Thus $S \subseteq \langle S \rangle$.
(ii) Let $x \in \langle S_1 \rangle$. Then there exists $s \in S_1$ such that $x \preceq s$. Since $s \in S_2, x \in \langle S_2 \rangle$.
(iii) By (1), $\langle S \rangle \subseteq \langle\langle S \rangle\rangle$. Let $x \in \langle\langle S \rangle\rangle$. Then there exists $y \in \langle S \rangle$ such that $x \preceq y$. There exists $s \in S$ such that $y \preceq s$. Since $\preceq$ is transitive, $x \preceq s$. Thus $x \in \langle S \rangle$. Hence $\langle\langle S \rangle\rangle \subseteq \langle S \rangle$.
(iv) For all $s \in S$, $\langle\{s\}\rangle \subseteq \langle S \rangle$ by (2). Thus $\cup_{s \in S}\langle\{s\}\rangle \subseteq \langle S \rangle$. Let $x \in \langle S \rangle$. Then there exists $s \in S$ such that $x \preceq s$. Hence $x \in \langle\{s\}\rangle$ and so $x \in \cup_{s \in S}\langle\{s\}\rangle$. Therefore, $\langle S \rangle \subseteq \cup_{s \in S}\langle\{s\}\rangle$.

(v) Suppose $x \in \langle S \cup \{y\}\rangle$ and $x \notin \langle S \rangle$. Then there does not exist $s \in S$ such that $x \preceq s$. Hence $x \preceq y$. Thus $x \in \langle \{y\}\rangle$.

Theorem 5.11 $\langle X \rangle = \langle PS_N(R)\rangle$.

Proof Clearly, $PS_N(R) \subseteq X$. Thus $\langle PS_N(R)\rangle \subseteq \langle X \rangle$. Let $x \in X$. If $x \notin \langle PS_N(R)\rangle$, then $x \notin PS_N(R)$ and so by (2) of Corollary 5.8, there exists $y \in PS_N(R)$ such that $x \prec y$. Thus $x \in \langle \{y\}\rangle \subseteq \langle PS_N(R)\rangle$. Hence $X \subseteq \langle PS_N(R)\rangle$ and so $\langle X \rangle \subseteq \langle PS_N(R)\rangle$.

Lemma 5.12

$$M(R, X) \cap PS_N(R) = \emptyset$$

if and only if $M(R, X) = \emptyset$.

Proof Suppose $M(R, X) \neq \emptyset$. Let $x \in M(R, X)$. By Theorem 5.11, there exists $y \in PS_N(R)$ such that $y \succeq x$. Since $x \in M(R, X)$, xRz for all $z \in X$. Since $y \succeq x$, yRz for all $z \in X$ by (P2). Thus $y \in M(R, X)$. Hence $M(R, X) \cap PS_N(R) \neq \emptyset$.

Lemma 5.13 *Let* $s \in PS_N(R)$. *Then there does not exist* $c \in PS_N(R)$ *such that* cPs *if and only if* $s \in M(R, X)$.

Proof Since R is complete and not cPs for all $c \in PS_N(R)$, it follows that sRc for all $c \in PS_N(R)$. Let $x \in X$. By Theorem 5.11, there exists $c \in PS_N(R)$ such that $c \succeq x$. Thus sRx by (P5). Hence $s \in M(R, X)$. The converse is immediate.

Lemma 5.14 (i) *Let* $s \in PS_N(R)$. *If there exists* $x \in X$ *such that* xPs, *then there exists* $c \in PS_N(R)$ *such that* cPs.
(ii) $M(R, X) = \emptyset$ *if and only if* $\forall s \in PS_N(R)$, *there exists* $c \in PS_N(R)$ *such that* cPs.

Proof (i) By Theorem 5.11, there exists $c \in PS_N(R)$ such that $c \succeq x$. Hence cPs by (P3).

(ii) Suppose $M(R, X) = \emptyset$. Then the result holds by Lemma 5.13. Conversely, suppose $M(R, X) \neq \emptyset$. By Lemma 5.12, $M(R, X) \cap PS_N(R) \neq \emptyset$ and so there exists $s \in M(R, X) \cap PS_N(R)$. Hence there does not exist $c \in PS_N(R)$ such that cPs. Let $V = \{v \in U \mid v$ is not in a cycle$\}$. Let $N_1 = V \backslash N_2$, where $N_2 = \{w \in V \mid \forall R \in \mathscr{R}, w \in PS_N(R) \Rightarrow M(R, X) \neq \emptyset\}$. Let $M_1 = \{w \in V \mid \forall R \in \mathscr{R}^n, w \notin PS_N(R)\}$. Assume $M_1 \subseteq N_1$. Let $N_1' = N_1 \backslash M_1$. Suppose N_1 is such that none of its elements are strictly preferred to one of $U \backslash V$.

Theorem 5.15 $M(R, X) = \emptyset$ *if and only if*

$$PS_N(R) = \left(\bigcup_{k=1}^{n} C_k \right) \cup \left(\bigcup_{j=1}^{m} C_j' \right) \cup N_1'',$$

where $N_1'' \subseteq N_1'$, C_k are cycles, $k = 1, \ldots, n$, C_j' subsets of cycles which are not themselves cycles, $j = 1, \ldots, m$, and

(i) $\forall s \in \bigcup_{j=1}^m C_j'$, there exists $c \in (\bigcup_{k=1}^n C_k) \cup (\bigcup_{j=1}^m C_j')$ such that cPs,

(ii) $\forall s \in N_1''$ there exists $c \in (\bigcup_{k=1}^n C_k) \cup (\bigcup_{j=1}^m C_j')$ such that cPs.

Proof It follows that $PS_N(R) \subseteq (\bigcup_{k=1}^n C_k) \cup (\bigcup_{j=1}^m C_j') \cup V$. Since no element of M_1 can be in $PS_N(R)$,

$$PS_N(R) \subseteq \left(\bigcup_{k=1}^n C_k \right) \cup \left(\bigcup_{j=1}^m C_j' \right) \cup (N_1 \backslash M_1) \cup N_2.$$

Hence it follows that

$$PS_N(R) = \left(\bigcup_{k=1}^n C_k \right) \cup \left(\bigcup_{j=1}^m C_j' \right) \cup N_1'' \cup N_2'$$

for certain cycles $C_k, k = 1, \ldots, n$, C_j' subsets of cycles which are not themselves cycles, $j = 1, \ldots, m$, and for some $N_1'' \subseteq N_1'$, and $N_2' \subseteq N_2$.

Suppose $M(R, X) = \emptyset$. Since $N_2 \cap PS_N(R) \neq \emptyset$ implies $M(R, X) \neq \emptyset$,

$$PS_N(R) = \left(\bigcup_{k=1}^n C_k \right) \cup \left(\bigcup_{j=1}^m C_j' \right) \cup N_1'',$$

i.e., $N_2' = \emptyset$. Since no element of $N_1 \backslash M_1$ is preferred to one of $U \backslash V$, no element of $N_1 \backslash M_1$ is preferred to one of $PS_N(R)$. Hence $\forall s \in \bigcup_{j=1}^m C_j'$,

$$\exists c \in \left(\bigcup_{k=1}^n C_k \right) \cup \left(\bigcup_{j=1}^m C_j' \right)$$

such that cPs by Lemma 5.13 else $M(R, X) \neq \emptyset$. By Lemma 5.13, $\forall s \in N_1''$, there exists

$$c \in \left(\bigcup_{k=1}^n C_k \right) \cup \left(\bigcup_{j=1}^m C_j' \right)$$

such that cPs.

For the converse, the conditions imply $\forall s \in PS_N(R)$, $\exists c \in PS_N(R)$ such that cPs. Hence no element of $PS_N(R)$ is in $M(R, X)$. Thus by Lemma 5.12, $M(R, X) = \emptyset$.

Theorem 5.16
$$M(R, X) \subseteq \langle M(R, X) \cap PS_N(R) \rangle .$$

Proof If $M(R, X) = \emptyset$, then the result is immediate since $\langle \emptyset \rangle = \emptyset$. Suppose $M(R, X) \neq \emptyset$. Let $s \in M(R, X)$. Suppose $s \notin PS_N(R)$. Since $\langle X \rangle = \langle PS_N(R) \rangle$, $s \in \langle PS_N(R) \rangle$. Hence there exists $c \in PS_N(R)$ such that $s \prec c$. Since $sRx \; \forall x \in X$, $cRx \; \forall x \in X$. Thus $c \in M(R, X)$. Hence $c \in M(R, X) \cap PS_N(R)$. Thus $s \in \langle M(R, X) \cap PS_N)(R) \rangle$. Therefore,

$$M(R, X) \subseteq \langle M(R, X) \cap PS_N(R) \rangle .$$

The following examples show that both $M(R, X) \not\subseteq PS_N(R)$ and $M(R, X) \subseteq PS_N(R)$ are possible.

Example 5.17 Let $n = 3$ and $U = T^3$. Let $PS_N(R) = C \cup \{(0.75, 0.75, 0.75)\}$, where $C = \{(1, 0.5, 0), (0.5, 0, 1), (0, 1, 0.5)\}$. Then C is a cycle. Thus $M(R, X) \cap PS_N(R) = \{(0.75, 0.75, 0.75)\}$. It is easily verified that

$$M(R, X) = \{(0.75, 0.75, 0.75), (0.75, 0.75, 0.5), (0.75, 0.5, 0.75), (0.5, 0.75, 0.75)\}.$$

Here $M(R, X) \not\subseteq PS_N(R)$.

Example 5.18 Let $n = 3$ and $U = T^3$. Let $PS_N(R) = C \cup \{(0.75, 0.75, 0.75)\}$, where $C = \{(1, 0.75, 0), (0.75, 0, 1), (0, 1, 0.75)\}$. Then C is a cycle. Thus $M(R, X) \cap PS_N(R) = \{(0.75, 0.75, 0.75)\}$. It is easily verified that $M(R, X) = \{(0.75, 0.75, 0.75\}$. In this example, $M(R, X) \subseteq PS_N(R)$.

The use of fuzzy discrete preferences introduced earlier in the book permitted us in this chapter to examine the effect of indifference on the existence of a maximal set in an n-dimensional public choice model. We demonstrated that the conditions for avoiding intransitivity and thereby assuring a nonempty maximal set are far less restrictive in fuzzy multi-dimensional public choice models than they are for the conventional approach. In all but a limited number of cases, the maximal set is empty if and only if Pareto set contains a union of cycles.

Before we turn to an application of these concepts in a fuzzy multi-dimensional public choice model, we give consideration to an alternative solution set, the uncovered set. Since the uncovered set imposes the condition that players make decisions using an amendment agenda, we do not use it in our public choice models of government formation processes. (These processes do not follow a formal amendment agenda.) We include the materials for those readers who may wish to consider using the uncovered set in an appropriate institutional setting. The discussion largely follows that in Mordeson et al. (2010, 2011).

5.3   The Existence of a Majority Rule Maximal Set in Arbitrary n-Dimensional Public Choice Models

Gibilisco et al. (2014) demonstrates that if the number of actors is limited to three, $n = 3$, then the maximal set is empty in fuzzy public choice models when the Pareto set contains a union of cycles. We present the argument found in Mordeson and Clark (2010) extending the result to an arbitrary number of players in multi-dimensional public choice models. (The case for four players, $n = 4$, is a special situation.) We also formally identify the exceptions, which turn out to be relatively trivial from an empirical perspective.

The approach taken in this section is quite constructive. Let $N = \{1, \ldots, n\}$ denote a finite set of players and X denote a set of alternatives. If R is a binary relation on X, we let P denote the strict preference relation associated with R, i.e., $P = \{(x, y) \in R \mid (y, x) \notin R\}$. Let $\mathscr{R}$ denote the set of all binary relations on X that are reflexive, complete, and transitive. Let

$$\mathscr{R}^n = \{\rho \mid \rho = (R_1, \ldots, R_n), R_i \in \mathscr{R}, i = 1, \ldots, n\}.$$

Let $\rho \in \mathscr{R}^n$. Then the Pareto set of ρ is defined to be

$$PS_N(\rho) = \{x \in X \mid \forall y \in Y \; \exists i \in N, \; yP_ix \Rightarrow \exists j \in N, \; xP_jy\} \,.$$

If $R \in \mathscr{R}$, the maximal set of R with respect to a subset S of X is defined to be

$$M(R, S) = \{x \in S \mid \forall y \in S, \; xRy\}.$$

Define the binary relation R on X by $\forall x, y \in X, (x, y) \in R$ if and only if $|\{i \in N \mid xR_iy\}| \geq n/2$. Let $R(x, y; \rho) = \{i \in N \mid xR_iy\}$ and $P(x, y; \rho) = \{i \in N \mid xP_iy\}$. Then $(x, y) \in P$ if and only if $|P(x, y; \rho)| > n/2$. For this R, $M(R, X)$ is called the *majority rule maximal set*. An *aggregation rule* is a function from $\mathscr{R}^n$ into $\mathscr{B}$, where $\mathscr{B}$ is the set of all binary relations on X which are reflexive and complete.

Let $n \in \mathbb{N}$, the positive integers, and let $T = \{0, 0.25, 0.5, 0.75, 1\}$. Define the binary relation $\preceq$ on T^n by $\forall (a_1, \ldots, a_n), (b_1, \ldots, b_n) \in T^n$, we have that $(a_1, \ldots, a_n) \preceq (b_1, \ldots, b_n)$ if and only if $a_i \leq b_i$ for $i = 1, \ldots, n$. Then $\preceq$ is a partial order on T^n and $\preceq$ satisfies the properties listed in Sect. 5.2. Let $\bar{a} = (a_1, \ldots, a_n) \in T^n$.

Lemma 5.19   *If $(a_1, \ldots, a_n)P \cdots P(b_1, \ldots, b_n)$ is a cycle, then $(1 - b_1, \ldots, 1 - b_n)P \cdots P(1 - a_1, \ldots, 1 - a_n)$ is a cycle.*

Lemma 5.20   *Let $n = 2q + 1$, where $q \in \mathbb{N}$. Let $N_1 = \{\bar{a} \in T^n \mid a_i \in \{0, 0.25\}, i = 1, \ldots, n\} \cup \{\bar{a} \in T^n \mid q + 1$ or more of the a_i equal $0\}$ and $N_2 = \{\bar{a} \in T^n \mid a_i \in \{1, 0.75\}, i = 1, \ldots, n\} \cup \{\bar{a} \in T^n \mid q + 1$ or more of the a_i equal $1\}$. Then $\bar{a}$ is not in a cycle if and only if $\bar{a} \in N_1 \cup N_2$.*

Proof Let $\bar{a} \in N_2$. Suppose $a_i \in \{1, 0.75\}, i = 1, \ldots, n$. If there exists $\bar{b} = (b_1, \ldots, b_n) \in T^n$ such that $\bar{b} P \bar{a}$. Then $q + 1$ of $b_1, \ldots, b_n$ are greater than $q + 1$ of the corresponding $a_1, \ldots, a_n$, say $b_1 > a_1, \ldots, b_{q+1} > a_{q+1}$. Then $b_1 = \ldots = b_{q+1} = 1$ since $a_i \in \{1, 0.75\}, i = 1, \ldots, n$. Hence $\nexists \bar{c} \in T^n$ such that $\bar{c} P \bar{b}$. Thus $\bar{a}$ is not in a cycle. Clearly if $q + 1$ or more of the a_i are equal to 1, then $\bar{a}$ is not in a cycle. By Lemma 5.19, no element of N_1 is in a cycle.

Suppose $\bar{a} \notin N_1 \cup N_2$. We consider $\bar{a}$ of the form:

(1) $(1, \overset{q-r}{\ldots}, 1, 0.75, \overset{q+s}{\ldots}, 0.75, a_1, \ldots, a_{r-s+1})$, where there are $q - r$ 1s, $q + s$ 0.75s, $r, s = 0, 1, \ldots, q$; $2q - r + s \leq 2q$, i.e., $s \leq r$, and $a_1, \ldots, a_{r-s+1} < 0.75$.

(2) $(1, \overset{q-r}{\ldots}, 1, 0.75, \overset{q-s}{\ldots}, 0.75, a_1, \ldots, a_{r+s+1})$, where there are $q - r$ 1s, $q - s$ 0.75s, $r, s = 0, 1, \ldots, q$; and $a_1, \ldots, a_{r+s+1} < 0.75$.

Suppose $\bar{a}$ has the form in (1). Since $q - r + s \geq 0, 2q - r + s \geq q$. Suppose $2q - r + s \geq q + 1$. (The case $2q - r + s = q$ is considered separately.) Then

$$\left(1, \overset{q-r}{\ldots}, 1, 0.75, \overset{q+s}{\ldots}, 0.75, a_1, \ldots, a_{r-s+1}\right)$$
$$P$$
$$\left(0.5, \overset{q-r}{\ldots}, 0.5, 0.5, \overset{q+s}{\ldots}, 0.5, 1, \overset{r-s+1}{\ldots}, 1\right)$$
$$P$$
$$\left(0, \overset{q-r}{\ldots}, 0, 0, \overset{s}{\ldots}, 0, 1, \overset{q}{\ldots}, 1, 0.75, \overset{r-s+1}{\ldots}, 0.75\right)$$

is a cycle since $2q - r + s \geq q + 1, q - r + s + r - s + 1 = q + 1$, and $q + r - s + 1 \geq q + 1$ (since $s \leq r$). We now consider the case $2q - r + s = q$. Then $r - s = q$. This is equivalent to $r = q$ and $s = 0$. We have

$$\left(0.75, \overset{q}{\ldots}, 0.75, a_1, \ldots, a_{q+1}\right)$$
$$P$$
$$\left(0.5, \overset{q}{\ldots}, 0.50, 1, \overset{q}{\ldots}, 1\right)$$
$$P$$
$$\left(1, \overset{q-1}{\ldots}, 1, 0.25, 0.75, \overset{q+1}{\ldots}, 0.75\right)$$

is a cycle since one of the $a_i > 0$, say $a_1 > 0$.

Suppose $\bar{a}$ has the form in (2). Suppose $r + s + 1 - q \geq 0$ and $r \leq q - 1$. We consider the cases $r + s + 1 - q < 0$ and $r > q - 1$ separately. Note that $r + s + 1 - q < 0$ and $r > q - 1$ is impossible else $0 > r + s + 1 - q > q - 1 + s + 1 - q = s$. We have

$$\left(1, \overset{q-r}{\ldots}, 1, 0.75, \overset{q-s}{\ldots}, 0.75, a_1, \ldots, a_{r+s-q+1}, a_{r+s-q+1}, \ldots, a_{r+s+1}\right)$$

$$P$$

$$\left(1, \overset{q-r}{\ldots}, 1, 0.75, \overset{q-s}{\ldots}, 0.75, 0, \overset{r+s-q+1}{\ldots}, 0, 1, \overset{q}{\ldots}, 1\right)$$

$$P$$

$$\left(0, \overset{q-r}{\ldots}, 0, 1, \overset{q-s}{\ldots}, 1, 0.75, \overset{r+s+1}{\ldots}, 0.75\right)$$

is a cycle since $q - r + q - s + r + s - q + 1 = q + 1$ $(r + s + 1 - q \geq 0$ and $a_1, \ldots, a_{r+s-q+1} > 0$ since q or more of the a_i cannot equal 0$)$, $q - r + q \geq q + 1$ $(r \leq q - 1)$, and $q - s + r + s + 1 \geq q + 1$ (the $a_i < 0.75$).

Suppose $r > q - 1$, i.e., $r = q$. Then

$$\left(0.75, \overset{q-s}{\ldots}, 0.75, a_1, \ldots, a_{q+s+1}\right)$$

$$P$$

$$\left(0.5, \overset{q-s}{\ldots}, 0.5, 0, \overset{s+1}{\ldots}, 0, 1, \overset{q}{\ldots}, 1\right)$$

$$P$$

$$\left(0, \overset{q-s}{\ldots}, 0, 1, \overset{s+1}{\ldots}, 1, 0.75, \overset{q}{\ldots}, 0.75\right)$$

is a cycle since $s + 1$ of the $a_1, \ldots, a_{q+s+1} > 0$, say $a_1, \ldots, a_{s+1}, q - s + q \geq q + 1$, i.e., $s \leq q - 1$ (the case $s = q$ is special and is considered next), and $s + 1 + q \geq q + 1$. (Recall that $a_i < 0.75$).

Now suppose $s = q$. Then we consider $(a_1, \ldots, a_{2q+1})$. That is, no components are equal to 0.75 or 1. By Lemma 5.19, no elements are equal to 0.25 or 0. Thus we have

$$\left(0.5, \overset{2q+1}{\ldots}, 0.5\right)$$

$$P$$

$$\left(0, \overset{q}{\ldots}, 0, 0.25, 1, \overset{q}{\ldots}, 1\right)$$

$$P$$

$$\left(1, \overset{q}{\ldots}, 1, 0, 0.75, \overset{q}{\ldots}, 0.75\right)$$

is a cycle.

The case $r + s + 1 - q < 0$ remains. Then $q - r - s > 1$, i.e., $2q - r - s > q + 1$. Since $r + s + 1 - q < 0$, $q - r > s + 1$ and $q - s > r + 1$. Hence

$$\left(1, \overset{q-r}{\ldots}, 1, 0.75, \overset{q-s}{\ldots}, 0.75, a_1, \ldots, a_{r+s+1}\right)$$

$$P$$

$$\left(0.5, \ldots, 0.5, 1, \overset{r+s+1}{\ldots}, 1\right)$$

$$P$$

$$\left(0, \ldots, 0, 1, \ldots, 1, 0.75, \overset{q}{\ldots}, 0.75\right)$$

is a cycle.

Unless otherwise specified, we assume $q \geq 3$ in the following.

Lemma 5.21 *Let $n = 2q$. Let $N_1 = \{\bar{a} \in T^n \mid n-1 \text{ or more of the } a_i \in \{0, 0.25\}, i = 1, \ldots, n\} \cup \{\bar{a} \in T^n \mid q \text{ or more of the } a_i \text{ equal } 0\}$ and $N_2 = \{\bar{a} \in T^n \mid n-1 \text{ or more of the } a_i \in \{1, 0.75\}, i = 1, \ldots, n\} \cup \{\bar{a} \in T^n \mid q \text{ or more of the } a_i \text{ equal } 1\}$. Then $\bar{a}$ is not in a cycle if and only if $\bar{a} \in N_1 \cup N_2$.*

Proof Let $\bar{a} \in N_2$. Suppose $n - 1$ or more of the $a_i \in \{1, 0.75\}$. If there exists $\bar{b} = (b_1, \ldots, b_n) \in T^n$ such that $\bar{b}P\bar{a}$. Then $q+1$ of $b_1, \ldots, b_n$ are greater than $q+1$ of the corresponding $a_1, \ldots, a_n$. Thus q or more of the b_i equal 1. Hence $\not\exists \bar{c} \in T^n$ such that $\bar{c}P\bar{b}$. Thus $\bar{a}$ is not in a cycle. Clearly, if q or more of the a_i are equal to 1, then $\bar{a}$ is not in a cycle. By Lemma 5.19, no element of N_1 is in a cycle.

Suppose $\bar{a} \notin N_1 \cup N_2$. We consider $\bar{a}$ of the form:

(1) $(1, \overset{q-r}{\ldots}, 1, 0.75, \overset{q+s}{\ldots}, 0.75, a_1, \ldots, a_{r-s})$, where there are $q-r$ 1s and $q+s$ 0.75s, $r, s = 0, 1, \ldots, q$; $2q - r + s \leq 2q$, i.e., $s \leq r$, and $a_1, \ldots, a_{r-s} < 0.75$.

(2) $(1, \overset{q-r}{\ldots}, 1, 0.75, \overset{q-s}{\ldots}, 0.75, a_1, \ldots, a_{r+s})$, where there are $q-r$ 1s, $q-s$, 0.75s, $r, s = 0, 1, \ldots, q$; and $a_1, \ldots, a_{r+s} < 0.75$.

Suppose $\bar{a}$ is of the form given in (1). Then $q - r + q + s \geq 2$ i.e., $2q - r + s \geq 2$. Also, $r \geq 1$ else $\bar{a} \in N_2$. Suppose $q - r + q + s \geq q + 1$. Then

$$\left(1, \ldots, 1, 0.75, \overset{q+s}{\ldots}, 0.75, a_1, \ldots, a_{r-s}\right)$$

$$P$$

$$\left(0.5, \overset{q+1}{\ldots}, 0.5, 1, \overset{q-1}{\ldots}, 1\right)$$

$$P$$

$$\left(0, \overset{q-1}{\ldots}, 0, 1, 1, 0.75, \overset{q-1}{\ldots}, 0.75\right)$$

is a cycle since $2q - 2 \geq q + 1(q \geq 3)$. Suppose $q - r + q + s \leq q$. Then $-q + 1 + r - s \geq 1$ and recall that $q - r + q + s \geq 2$. Thus

$$\left(1,\overset{q-r}{\ldots},1,0.75,\overset{q+s}{\ldots},0.75,a_1,\ldots,a_{-q+1+r-s},a_{-q+2+r-s},\overset{q-1}{\ldots},a_{r-s}\right)$$

$$P$$

$$\left(0.5,\overset{q-r}{\ldots},0.5,0.5,\overset{q+s}{\ldots},0.5,\overset{-q+1+r-s}{0,\ldots,0},1,\overset{q-1}{\ldots},1\right)$$

$$P$$

$$\left(0,\overset{q-r}{\ldots},0,0,\overset{q+s}{\ldots},0\overset{-q+1+r-s}{1,\ldots,1},0.75,\overset{q-1}{\ldots},0.75)\right)$$

is a cycle since $a_1,\ldots,a_{-q+1+r-s} > 0$ and if we assume $-q+1+r-s \geq 2$.
Suppose $-q+1+r-s = 1$. Then $r-s = q$ and so

$$\left(1,\overset{q-r}{\ldots},1,0.75,\overset{q+s}{\ldots},0.75,a_1,a_2,\overset{q-1}{\ldots},a_{r-s}\right)$$

$$P$$

$$\left(0.5,\overset{q-r}{\ldots},0.5,0.5,\overset{q+s}{\ldots},0.5,0,1,\overset{q-1}{\ldots},1\right)$$

$$P$$

$$\left(0,\overset{q-1}{\ldots},0,1,\overset{q+s-1}{\ldots},11,0.75,\overset{q-1}{\ldots},0.75)\right)$$

is a cycle since $a_1 > 0$, $2q-2 \geq q+1$ ($q \geq 3$), and $q-s+1 \geq 1$.

Suppose $\bar{a}$ has the form given in (2). Suppose $r+s \geq q+1$. Recall $q-r+q-s \geq 2$. Then

$$\left(1,\overset{q-r}{\ldots},1,0.75,\overset{q-s}{\ldots},0.75,a_1,\ldots,a_{-q+1+r+s},a_{-q+2+r+s},\overset{q-1}{\ldots},a_{r+s}\right)$$

$$P$$

$$\left(0.5,\overset{q-r}{\ldots},0.5,0.5,\overset{q-s}{\ldots},0.5,\overset{-q+1+r+s}{0,\ldots,0},1,\overset{q-1}{\ldots},1\right)$$

$$P$$

$$\left(0,\overset{q-r}{\ldots},0,0,\overset{q-s}{\ldots},0,\overset{-q+1+r+s}{1,\ldots,1},0.75,\overset{q-1}{\ldots},0.75\right)$$

is a cycle since $a_1,\ldots,a_{-q+1+r+s} > 0$, $-q+1+r+s \geq 2$ and $a_{-q+2+r+s},\ldots,$
$a_{r+s} < 0.75$. We now consider the case $r+s < q+1$. Suppose $r+s = q$. Then

$$\left(1,\overset{q-r}{\ldots},1,0.75,\overset{q-s}{\ldots},0.75,a_1,a_2,\overset{q-1}{\ldots},a_q\right)$$

$$P$$

$$\left(0.5,\overset{q-r}{\ldots},0.5,0.5,\overset{q-s}{\ldots},0.5,0,1,\overset{q-1}{\ldots},1\right)$$

$$P$$

$$\left(0,\overset{q-1}{\ldots},0,1,1,0.75,\overset{q-1}{\ldots},0.75\right)$$

is a cycle since $a_1 > 0, a_2, \ldots, a_q < 0.75$, and $q - s \geq 1$ ($r \geq 1$ and $r + s = q$). Suppose $r + s < q$. Recall $r + s \geq 2$.

$$\left(1, \overset{q-r}{\ldots}, 1, 0.75, \overset{q-s}{\ldots}, 0.75, a_1, \ldots, a_{r+s} \right)$$

$$P$$

$$\left(0.5, \overset{q+1}{\ldots}, 0.5, 1, \overset{q-1}{\ldots}, 1 \right)$$

$$P$$

$$\left(0, \overset{q-1}{\ldots}, 0, 1, \overset{2}{\ldots}, 1, 0.75, \overset{q-1}{\ldots}, 0.75 \right)$$

is a cycle since $q - r + q - s \geq q + 1, 2q - 2 \geq q + 1$ ($q \geq 3$), and $q + r \geq q + 1$ ($r \geq 1$).

We now consider the case $q = 2$. Let $n = 2q = 4$.

Lemma 5.22 T^4 *has no 3-cycles.*

Proof Suppose T^4 has a 3-cycle, say

$$(a_1, a_2, a_3, a_4)$$
$$P$$
$$(b_1, b_2, b_3, b_4)$$
$$P$$
$$(c_1, c_2, c_3, c_4) \ .$$

Then three of the a_i are greater than three of the corresponding b_i, say $a_1 > b_1$, $a_2 > b_2$, and $a_3 > b_3$. Also three of the b_i are greater than three of the corresponding c_i, say (1) $b_4 \leq c_4$ so $b_1 > c_1, b_2 > c_2, b_3 > c_3$ or (2) $b_4 > c_4$. Since three of the c_i are greater than three of the corresponding a_i, it follows in either case (1) or (2) that $\exists i$ such that $a_i > b_i > c_i > a_i$ which is impossible.

Note that if $a_1 > a_2 > a_3 > a_4$, then

$$(a_1, a_2, a_3, a_4)$$
$$P$$
$$(a_2, a_3, a_4, a_1)$$
$$P$$
$$(a_3, a_4, a_1, a_2)$$
$$P$$
$$(a_4, a_1, a_2, a_3)$$

is a 4-cycle.

Lemma 5.23 *Let $q = 2$, i.e., $n = 4$. Let $N_1 = \{\bar{a} \in T^4 \mid$ three or more of the $a_i \in \{0, 0.25\}$, $i = 1, 2, 3, 4\} \cup \{\bar{a} \in T^4 \mid$ two or more of the $a_i = 1\}$, $N_2 = \{\bar{a} \in T^4 \mid$ three or more of the $a_i \in \{1, 0.75\} \cup \{\bar{a} \in T^4 \mid$ two or more of the $a_i = 1\}$, $N_3 = \{\bar{a} \in T^4 \mid$ three or more of the $a_i = 0.5\}$, and $N_4 = \{\bar{a} \in T^4 \mid$ two of the $a_i = 0.5$ and two of the $a_i \in \{1, 0.75\}$ or two of the $a_i \in \{0, 0.25\}$. Then $\bar{a}$ is not in a cycle if and only if $\bar{a} \in N_1 \cup N_2 \cup N_3 \cup N_4$.*

Proof Suppose $\bar{a} \in N_2$. If two or more of the $a_i = 1$, then $\nexists \bar{b} \in T^4$ such that $\bar{b} P \bar{a}$. Hence $\bar{a}$ is not in a cycle. Suppose three or more of the $a_i \in \{1, 0.75\}$. If $\bar{b} P \bar{a}$ for some $\bar{b} \in T^4$, then two or more of the b_i equal 1 and so $\bar{b}$ and hence $\bar{a}$ is not in a cycle. Suppose $\bar{a} \in N_3$, say $\bar{a} = (0.5, 0.5, 0.5, a)$, $a \in \{1, 0.75\}$. If $\bar{b} P \bar{a}$ for some $\bar{a} \in T^4$, then three or more of the $b_i \geq 0.75$ and so $\bar{b}$ and hence $\bar{a}$ is not in a cycle. (If $a \in \{0, 0.25\}$, then $\bar{a}$ is not in a cycle by Lemma 5.19) If $\bar{a} \in N_1$, then by Lemma 5.19 and the above argument $\bar{a}$ is not in a cycle. Suppose $\bar{a} \in N_4$. If two of the $a_i \in \{1, 0.75\}$ and $\bar{b} P \bar{a}$ for some $\bar{a} \in T^4$, the three of the $b_i \geq 0.75$ so $\bar{b}$ and hence $\bar{a}$ is not in a cycle. (If two of the $a_i \in \{0, 0.25\}$, then $\bar{a}$ is not in a cycle by Lemma 5.19)

Suppose $\bar{a} \notin N_1 \cup N_2 \cup N_3 \cup N_4$. Then by Lemma 5.19 it suffices to show the following $\bar{a}$ are in a cycle.

(1) $(1, 0.75, a_1, a_2)$, where $a_1 < 0.5$ or $a_2 < 0.5$ and $a_1 > 0$ or $a_2 > 0$ and $a_1 < 0.75, a_2 < 0.75$;
(2) $(0.75, 0.75, a_1, a_2)$, where $a_1 < 0.5$ or $a_2 < 0.5$ and $a_1 > 0$ or $a_2 > 0$ and $a_1 < 0.75, a_2 < 0.75$;
(3) $(1, a_1, a_2, a_3)$, where $a_1, a_2, a_3 < 0.75$ and at least two of $a_1, a_2, a_3 > 0$;
(4) $(0.75, a_1, a_2, a_3)$, where $a_1, a_2, a_3 < 0.75$ and at least two of $a_1, a_2, a_3 > 0$;
(5) $(0.5, 0.5, a_1, a_2)$, where not both a_1, a_2 are in $\{0, 0.25\}$ and not both a_1, a_2 are in $\{1, 0.75\}$ and $a_1 \neq 0.5 \neq a_2$.

Suppose $\bar{a}$ is of the form given in (1). Then by comments immediately preceding the lemma, it suffices to consider the case $a_1 = a_2 = 0.25$. Thus

$$(1, 0.75, 0.25, 0.25) \, P \, (0.75, 0.5, 0, 1) \, P \, (0.5, 0, 1, 0.75) \, P \, (0, 1, 0.75, 0.5)$$

is a cycle.

Suppose $\bar{a}$ is of the form given in (2). Suppose $a_1 > 0$. Then if $a_2 < 0.5$

$$(0.75, 0.75, a_1, a_2) \, P \, (0.5, 0.5, 0, 1) \, P \, (0.25, 0, 1, 0.75) \, P \, (0, 1, 0.75, 0.5)$$

is a cycle. Suppose $a_2 = 0.5$. Then $a_1 = 0.25$ else $\bar{a} \in N_4$. Thus

$$(0.75, 0.75, 0.25, 0.5) \, P \, (0.5, 0.5, 1, 0) \, P \, (0.25, 0, 0.75, 1) \, P \, (0, 1, 0.5, 0.75)$$

is a cycle.

Suppose $\bar{a}$ is of the form given in (3). Since all four of the a_i distinct has been covered and since $\bar{a} \notin N_1 \cup N_2 \cup N_3 \cup N_4$, the following cases remain.

$$(1, 0.5, 0.5, 0), (1, 0.5, 0.5, 0.25), (1, 0.5, 0.25, 0.25).$$

Now

$(1, 0.5, 0.5, 0) P. (75, 0.25, 0.25, 1) P. (25, 0, 1, 0.75) P (0, 0.75, 0.75, 0.5),$

$(1, 0.5, 0.5, 0.25) P (0.75, 0.25, 0.25, 1) P (0.25, 0, 1, 0.75) P (0, 0.75, 0.75, 0.5),$

$(1, 0.5, 0.25, 0.25) P (0.75, 0.25, 0, 1) P (0.25, 0, 1, 0.75) P (0, 1, 0.75, 0.5)$

are cycles.

Suppose $\bar{a}$ is of the form given in (4). Now for $a \in \{0, 0.25\}$,

$(0.75, 0.5, 0.5, a) P (0.5, 0.25, 0.25, 1) P (0.25, 0, 1, 0.75) P (0, 1, 0.75, 0.5),$

$(0.75, 0.5, 0.25, 0.25) P (0.5, 0.25, 0, 1) P (0.25, 0, 1, 0.75) P (0, 1, 0.5, 0.5)$

are cycles.

Suppose $\bar{a}$ has the form given in (5). The cases

$$(0.5, 0.5, 0, 1), (0.5, 0.5, 0.25, 1), (0.5, 0.5, 0, 0.75), (0.5, 0.5, 0.25, 0.75)$$

have been shown above to be in a cycle. Thus the proof is complete.

We next consider the case, where $T = \{0, 0.5, 1\}$.

We can now present our major result for an arbitrary number of players (except for $n = 4$). Here $I_1 = \{\bar{a} \in T^n \mid a_i \in \{0, 1\}, i = 1, \ldots, n \text{ and } \exists! a_i = 1\}$ and $N_1' = \{\bar{a} \in T^n \mid a_i \in \{0, 0.25\}, i = 1, \ldots, n \text{ and } \exists \text{unique } a_i = 0.25\}$.

Theorem 5.24 *Suppose $n \neq 4$. $M(R, X) = \emptyset$ if and only if $PS_N(R) = \left(\bigcup_{i=1}^{k} C_i\right) \cup \left(\bigcup_{j=1}^{m} C_j'\right) \cup N_1'' \cup I_1'$, where $N_1'' \subseteq N_1', I_1' \subseteq I_1, C_i$ are cycles, $i = 1, .., k, C_j'$ are subsets of cycles which are not themselves cycles, $j = 1, \ldots, m$, and*

(1) $\forall \bar{s} \in \bigcup_{j=1}^{m} C_j'$, there exists $\bar{c} \in \left(\bigcup_{i=1}^{k} C_i\right) \cup \left(\bigcup_{j=1}^{m} C_j'\right)$ such that $\bar{c} P \bar{s}$,

(2) $\forall \bar{s} \in N_1''$ there exists $\bar{c} \in \left(\bigcup_{i=1}^{k} C_i\right) \cup \left(\bigcup_{j=1}^{m} C_j'\right)$ such that $\bar{c} P \bar{s}$ and

(3) $\forall i \in I_1', \exists \bar{d} \in \left(\bigcup_{i=1}^{k} C_i\right) \cup \left(\bigcup_{j=1}^{m} C_j'\right) \cup N_1''$ such that $\bar{d} P \bar{i}$.

Proof By Lemmas 5.20 and 5.21, it follows that $PS_N(R) \subseteq \left(\bigcup_{i=1}^{k} C_i\right) \cup \left(\bigcup_{j=1}^{m} C_j'\right) \cup N_1 \cup N_2$. Since no element of $\langle I_1 \rangle \backslash I_1$ can be in $PS_N(R)$ and since no element of $N_1 \backslash N_1'$ can be in $PS_N(R), PS_N(R) \subseteq \left(\bigcup_{i=1}^{k} C_i\right) \cup \left(\bigcup_{j=1}^{m} C_j'\right) \cup N_1' \cup I_1 \cup N_2$. Hence it follows that $PS_N(R) = \left(\bigcup_{i=1}^{k} C_i\right) \cup \left(\bigcup_{j=1}^{m} C_j'\right) \cup N_1'' \cup I_1' \cup N_2'$ for certain cycles $C_i, i = 1, \ldots, k, C_j'$ subsets of cycles which are not themselves cycles, $j = 1, \ldots, m$, and for some $N_1'' \subseteq N_1', I_1' \subseteq I_1$, and $N_2' \subseteq N_2$.

Suppose $M(R, X) = \emptyset$. By Lemma 5.12, $PS_N(R) = \left(\bigcup_{i=1}^{k} C_i\right) \cup \left(\bigcup_{j=1}^{m} C_j'\right) \cup$
$N_1'' \cup I_1'$, i. e., $N_2' = \emptyset$. If $\bar{s} \in N_1$ is such that $\bar{s}P\bar{c}$ for some $\bar{c} \in PS_N(R)$, then $q + 1$
of the components of $\bar{c}$ equal 0 and so $\bar{c}$ is not in a cycle. Thus $\bar{c} \in N_1'' \cup I_1'$ which
is impossible. Hence no element of N_1 is strictly preferred to one of $\left(\bigcup_{i=1}^{k} C_i\right) \cup$
$\left(\bigcup_{j=1}^{m} C_j'\right)$. No element of I_1' is strictly preferred to any element of $PS_N(R)$. Thus $\forall \bar{s} \in$
$\bigcup_{j=1}^{m} C_j'$, $\exists \bar{c} \in \left(\bigcup_{i=1}^{k} C_i\right) \cup \left(\bigcup_{j=1}^{m} C_j'\right)$ such that $\bar{c}P\bar{s}$ by Lemma 5.13 else $M(R, X) \neq$
$\emptyset$. By Lemma 5.13, $\forall \bar{s} \in N_1''$ there exists $\bar{c} \in \left(\bigcup_{i=1}^{k} C_i\right) \cup \left(\bigcup_{j=1}^{m} C_j'\right)$
such that $\bar{c}P\bar{s}$ and $\forall i \in I_1'$, $\exists \bar{d} \in \left(\bigcup_{i=1}^{k} C_i\right) \cup \left(\bigcup_{j=1}^{m} C_j'\right) \cup N_1''$ such that $\bar{d}P\bar{i}$.

For the converse, the conditions imply $\forall \bar{s} \in PS_N(R)$, $\exists \bar{c} \in PS_N(R)$ such that $\bar{c}P\bar{s}$.
Hence no element of $PS_N(R)$ is in $M(X, R)$. Thus by Lemma 5.14, $M(R, X) = \emptyset$.

The following result addresses the special case when the number of players is
four.

Theorem 5.25 *Suppose $n = 4$. Then $M(R, X) = \emptyset$ if and only if*

$$PS_N(R) = \left(\bigcup_{i=1}^{k} C_i\right) \cup \left(\bigcup_{j=1}^{m} C_j'\right) \cup N_1'' \cup I_1' \cup N_3' \cup N_4',$$

where $N_1'' \subseteq N_1'$, $I_1' \subseteq I_1$, $N_3' \subseteq N_3$,

$$N_4' \subseteq N_4 \backslash \{\bar{a} \in T^4 \mid \text{two of the } a_i = 0.5 \text{ and two of the } a_i = 1\},$$

*C_i, are cycles, $i = 1, .., k$, C_j' are subsets of cycles which are not themselves cycles,
$j = 1, \ldots m$, and*

(1) $\forall \bar{s} \in \bigcup_{j=1}^{m} C_j'$, there exists $\bar{c} \in \left(\bigcup_{i=1}^{k} C_i\right) \cup \left(\bigcup_{j=1}^{m} C_j'\right) \cup N_3' \cup N_4'$ such that $\bar{c}P\bar{s}$,

(2) $\forall \bar{s} \in N_1''$ there exists $\bar{c} \in \left(\bigcup_{i=1}^{k} C_i\right) \cup \left(\bigcup_{j=1}^{m} C_j'\right) \cup N_3' \cup N_4'$ such that $\bar{c}P\bar{s}$,

(3) $\forall \bar{s} \in N_3' \cup N_4'$, there exists $\bar{c} \in \left(\bigcup_{i=1}^{k} C_i\right) \cup \left(\bigcup_{j=1}^{m} C_j'\right)$ such that $\bar{c}P\bar{s}$,

(4) $\forall i \in I_1'$, $\exists \bar{d} \in \left(\bigcup_{i=1}^{k} C_i\right) \cup \left(\bigcup_{j=1}^{m} C_j'\right) \cup N_1'' \cup N_3' \cup N_4'$ such that $\bar{d}P\bar{i}$.

Proof By Lemma 5.23, it follows that

$$PS_N(R) \subseteq \left(\bigcup_{i=1}^{k} C_i\right) \cup \left(\bigcup_{j=1}^{m} C_j'\right) \cup N_1 \cup N_2 \cup N_3 \cup N_4.$$

Since no element of $\langle I_1 \rangle \setminus I_1$ can be in $PS_N(R)$ and since no element of $N_1 \setminus N_1'$ can be in

$$PS_N(R), PS_N(R) \subseteq \left(\bigcup_{i=1}^{k} C_i \right) \cup \left(\bigcup_{j=1}^{m} C_j' \right) \cup N_1' \cup I_1 \cup N_2 \cup N_3 \cup N_4.$$

Hence it follows that

$$PS_N(R) = \left(\bigcup_{i=1}^{k} C_i \right) \cup \left(\bigcup_{j=1}^{m} C_j' \right) \cup N_1'' \cup I_1' \cup N_2' \cup N_3' \cup N_4'$$

for certain cycles $C_i, i = 1, \ldots, k$, C_j' subsets of cycles which are not themselves cycles, $j = 1, \ldots, m$, and for some

$$N_1'' \subseteq N_1', \ I_1' \subseteq I_1, \ N_2' \subseteq N_2, \ N_3' \cup N_4'$$

since $\{ \bar{a} \in T^4 \mid \text{two of the } a_i = 0.5 \text{ and two of the } a_i = 1 \} \subseteq N_2$.

Suppose $M(R, X) = \emptyset$. Then $M(R, X) \cap PS_N(R) = \emptyset$ and

$$PS_N(R) = \left(\bigcup_{i=1}^{k} C_i \right) \cup \left(\bigcup_{j=1}^{m} C_j' \right) \cup N_1'' \cup I_1' \cup N_3' \cup N_4',$$

i.e., $N_2' = \emptyset$. If $\bar{s} \in N_1$ is such that $\bar{s} P \bar{c}$ for some $\bar{c} \in PS_N(R)$, then 3 of the components of $\bar{c}$ equal 0 and so $\bar{c}$ is not in a cycle. Thus $\bar{c} \in N_1'' \cup I_1'$ which is impossible. Hence no element of N_1 is strictly preferred to one of $\left(\bigcup_{i=1}^{k} C_i \right) \cup \left(\bigcup_{j=1}^{m} C_j' \right)$. No element of I_1' is strictly preferred to any element of $PS_N(R)$. Clearly no element of $N_3 \cup N_4$ is strictly preferred to an element from $N_3 \cup N_4$. Thus

$$\forall \bar{s} \in \bigcup_{j=1}^{m} C_j', \exists \bar{c} \in \left(\bigcup_{i=1}^{k} C_i \right) \cup \left(\bigcup_{j=1}^{m} C_j' \right) \cup N_3' \cup N_4'$$

such that $\bar{c} P \bar{s}$ by Lemma 5.13 else $M(R, X) \neq \emptyset$. By Lemma 5.13, $\forall \bar{s} \in N_1''$ there exists

$$\bar{c} \in \left(\bigcup_{i=1}^{k} C_i \right) \cup \left(\bigcup_{j=1}^{m} C_j' \right) \cup N_3' \cup N_4'$$

such that $\overline{c}P\overline{s}$ and

$$\forall i \in I_1', \exists \overline{d} \in \left(\bigcup_{i=1}^{k} C_i \right) \cup \left(\bigcup_{j=1}^{m} C_j' \right) \cup N_1'' \cup N_3' \cup N_4'$$

such that $\overline{d}P\overline{i}$.

For the converse, the conditions imply $\forall \overline{s} \in PS_N(R)$, $\exists \overline{c} \in PS_N(R)$ such that $\overline{c}P\overline{s}$. Hence no element of $PS_N(R)$ is in $M(X, R)$. Thus by Lemma 5.14, $M(R, X) = \emptyset$.

We next consider the case, where $T = \{0, 0.5, 1\}$.

Proposition 5.26 *Suppose $T = \{0, 0.5, 1\}$. Then T^4 has no cycles.*

Proof Suppose $\overline{a}P\overline{b}$, where $\overline{a} = (a_1, a_2, a_3, a_4)$ and $\overline{b} = (b_1, b_2, b_3 b_4)$. We show $\nexists \overline{c} \in T^4$ such that $\overline{b}P\overline{c}$. If $\overline{a}$ is in a cycle, then $\exists \overline{d} \in T^4$ such that $\overline{d}P\overline{a}$. Hence only one a_i can equal 1, say $a_1 = 1$. Now three of the a_i are greater than three of the corresponding b_i, say (1) $a_1 > b_1, a_2 > b_2, a_3 > b_3$ or (2) $a_2 > b_2, a_3 > b_3, a_4 > b_4$. Suppose (1) holds. Then $a_2 = 0.5$ and $a_3 = 0.5$. Hence $b_2 = 0$ and $b_3 = 0$. Hence $\nexists \overline{c} \in T^4$ such that $\overline{b}P\overline{c}$. Thus $\overline{a}$ is not in a cycle. The proof for case (2) is similar. ∎

Proposition 5.27 *Suppose $T = \{0, 0.5, 1\}$. Let $n \in \mathbb{N}$. Then T^n has a cycle for $n = 3$ and $n \geq 5$.*

Proof Suppose $n = 3$. Then $(1, 0.5, 0)P(0.5, 0, 1)P(0, 1, 0.5)$ is a cycle.

Let $\overline{a}_1 = (1, 0.5, 0)$, $\overline{a}_2 = (0.5, 0, 1)$, and $\overline{a}_3 = (0, 1, 0.5)$.

Consider $(a1, 1, 0)$ as the five-tuple $(1, 0.5, 0, 1, 0)$, where $\overline{a}_1 = (1, 0.5, 0)$. Also $(\overline{a}_1, \overline{a}_1)$ is considered as the corresponding 6-tuple.

Let $n = 5$. Then $(\overline{a}_1, 1, 0)P(\overline{a}_2, 0, 1)P(\overline{a}_3, 0.5, 0.5)$ is a cycle. For $n = 6$, $(\overline{a}_1, \overline{a}_1)P(\overline{a}_2, \overline{a}_2)P(\overline{a}_3, \overline{a}_3)$ is a cycle.

In general, consider

$$3q, 3q + 1, 3q + 2 \text{ for } q = 1, 2, \ldots.$$

For $q = 1$, we have cases $n = 3, 4, 5$ above. It suffices to consider the cases

(1) $2q > 3q/2$,
(2) $2q > (3q + 1)/2$, and
(3) $2q + 1 > (3q + 2)/2$.

For case (1), we have $4q > 3q$ for all q. For case (2), we have $4q > 3q + 1$ for all q except $q = 1$, i.e., the case $n = 4$. For case (3), we have $4q + 1 > 3q + 1$ for all q. ∎

Proposition 5.28 *Let $T = \{0, 0.5, 1\}$. Let $n = 2q + 1$, where $q \in \mathbb{N}$. Let $N_1 = \{\overline{a} \in T^n \mid a_i \in \{0, 0.5\}, i = 1, \ldots, n\} \cup \{\overline{a} \in T^n \mid q + 1 \text{ or more of the } a_i \text{ equal } 0\}$ and $N_2 = \{\overline{a} \in T^n \mid a_i \in \{1, 0.5\}, i = 1, \ldots, n\} \cup \{\overline{a} \in T^n \mid q + 1 \text{ or more of the } a_i \text{ equal } 1\}$. Then $\overline{a}$ is not in a cycle if and only if $\overline{a} \in N_1 \cup N_2$.*

Proof Let $\bar{a} \in N_2$. Suppose $a_i \in \{1, 0.5\}, i = 1, \ldots, n$. If there exists $\bar{b} = (b_1, \ldots, b_n) \in T^n$ such that $\bar{b}P\bar{a}$. Then $q + 1$ of $b_1, \ldots, b_n$ are greater than $q + 1$ of the corresponding $a_1, \ldots, a_n$, say $b_1 > a_1, \ldots, b_{q+1} > a_{q+1}$. Then $b_1 = \ldots = b_{q+1} = 1$ since $a_i \in \{1, 0.5\}, i = 1, \ldots, n$. Hence $\nexists \bar{c} \in T^n$ such that $\bar{c}P\bar{b}$. Thus $\bar{a}$ is not in a cycle. Clearly if $q + 1$ or more of the a_i are equal to 1, then $\bar{a}$ is not in a cycle. By Lemma 5.20, no element of N_1 is in a cycle.

Suppose $\bar{a} \notin N_1 \cup N_2$. Then at least one of the $a_i = 1$ and at least one equals 0. We consider $\bar{a}$ of the form:

(1) $(\overset{q-r}{1, \ldots, 1}, \overset{q+s}{0.5, \ldots, 0.5}, a_1, \ldots, a_{r-s+1})$, where there are $q - r$ 1s, $q + s$, 0.5s, $r, s = 0, 1, \ldots, q; 2q - r + s \leq 2q$, i.e., $s \leq r$, and $a_1, \ldots, a_{r-s+1} < 0.5$.

(2) $(\overset{q-r}{1, \ldots, 1}, \overset{q-s}{0.5, \ldots, 0.5}, a_1, \ldots, a_{r+s+1})$, where there are $q - r$ 1s, $q - s$, 0.5s, $r, s = 0, 1, \ldots, q$; and $a_1, \ldots, a_{r+s+1} < 0.5$.

Suppose $\bar{a}$ is of the form given in 1. Then $r - s + 1 \geq 1$, i.e., $s \leq r$ and $q - r \geq 1$, i.e., $r \leq q - 1$. (Thus $s \leq q - 1$.) Hence

$$\left(\overset{q-r}{1, \ldots, 1}, \overset{q+s}{0.5, \ldots, 0.5}, \overset{r-s+1}{0, \ldots, 0} \right)$$

$$P$$

$$\left(\overset{q-r+s}{0.5, \ldots, 0.5}, \overset{q}{0, \ldots, 0}, \overset{r-s+1}{1, \ldots, 1} \right)$$

$$P$$

$$\left(\overset{q-r+s}{0, \ldots, 0}, \overset{q}{1, \ldots, 1}, \overset{r-s+1}{0.5, \ldots, 0.5} \right)$$

is a cycle since $q - r + q \geq q + 1$ since $q - r \geq 1$ else $\bar{a} \in N_1, q - r + s + r - s + 1 = q + 1$, and $q + r - s + 1 \geq q + 1$.

Suppose $\bar{a}$ has the form given in 2. Then

$$\left(\overset{q-r}{1, \ldots, 1}, \overset{q-s}{0.5, \ldots, 0.5}, \overset{r+s+1}{0, \ldots, 0} \right)$$

$$P$$

$$\left(\overset{q-r}{0.5, \ldots, 0.5}, \overset{q-s}{0, \ldots, 0}, \overset{r+s+1}{1, \ldots, 1} \right)$$

$$P$$

$$\left(\overset{q-r}{0, \ldots, 0}, \overset{q-s}{1, \ldots, 1}, \overset{r+s+1}{0.5, \ldots, 0.5} \right)$$

is a cycle since $q - r + q - s \geq q + 1$ else $r + s + 1 > q$ and so $\bar{a} \in N_1, q - r + r + s + 1 \geq q + 1$, and $q - s + r + s + 1 \geq q + 1$.

Proposition 5.29 *Let $T = \{0, 0.5, 1\}$. Let $n = 2q$. Suppose $q > 2$. Let $N_1 = \{\bar{a} \in T^n \mid n - 1$ or more of the $a_i \in \{0, 0.5\}, i = 1, \ldots, n\} \cup \{\bar{a} \in T^n \mid q$ or more of the a_i equal $0\}$ and $N_2 = \{\bar{a} \in T^n \mid n - 1$ or more of the $a_i \in \{1, 0.5\}, i = 1, \ldots, n\} \cup \{\bar{a} \in T^n \mid q$ or more of the a_i equal $1\}$. Then $\bar{a}$ is not in a cycle if and only if $\bar{a} \in N_1 \cup N_2$.*

Proof Let $\bar{a} \in N_2$. Suppose $n - 1$ or more of the $a_i \in \{1, 0.5\}$. If there exists $\bar{b} = (b_1, \ldots, b_n) \in T^n$ such that $\bar{b}P\bar{a}$. Then $q + 1$ of $b_1, \ldots, b_n$ are greater than $q + 1$ of the corresponding $a_1, \ldots, a_n$. Thus q or more of the b_i equal 1. Hence $\nexists \bar{c} \in T^n$ such that $\bar{c}P\bar{b}$. Thus $\bar{a}$ is not in a cycle. Clearly, if q or more of the a_i are equal to 1, then $\bar{a}$ is not in a cycle. By Lemma 5.21, no element of N_1 is in a cycle.

Suppose $\bar{a} \notin N_1 \cup N_2$. Then at least one of the $a_i = 1$ and at least one equals 0. We consider $\bar{a}$ of the form:

(1) $(1, \overset{q-r}{\ldots}, 1, 0.5, \overset{q+s}{\ldots}, 0.5, a_1, \ldots, a_{r-s})$, where there are $q - r$ 1s and $q + s$ 0.5s, $r, s = 0, 1, \ldots, q$; $2q - r + s \le 2q$, i.e., $s \le r$, and $a_1, \ldots, a_{r-s} < 0.5$.

(2) $(1, \overset{q-r}{\ldots}, 1, 0.5, \overset{q-s}{\ldots}, 0.5, a_1, \ldots, a_{r+s})$, where there are $q - r$ 1s, $q - s$ 0.5s, $r, s = 0, 1, \ldots, q$; and $a_1, \ldots, a_{r+s} < 0.5$.

Suppose $\bar{a}$ is of the form given in 1. Then $r - s \ge 1$, i.e., $s \le r - 1$ and $q - r \ge 1$, i.e., $r \le q - 1$. (Thus $s \le q - 2$.) Hence

$$\left(1, \overset{q-r}{\ldots}, 1, 0.5, \overset{q+s}{\ldots}, 0.5, 0, \overset{r-s}{\ldots}, 0\right)$$
$$P$$
$$\left(0.5, \overset{q-r}{\ldots}, 0.5, 0, \overset{q-1}{\ldots}, 0, 0.5, \overset{s+1}{\ldots}, 0.51, \overset{r-s}{\ldots}, 1\right)$$
$$P$$
$$\left(0, \overset{q-r}{\ldots}, 0, 1, \overset{q-1}{\ldots}, 1, 0, \overset{s+1}{\ldots}, 00.5, \overset{r-s}{\ldots}, 0.5\right)$$

is a cycle since $q - r + q - 1 \ge 2 + q - 1 = q + 1, q - r + s + 1 + r - s = q + 1$, and $q - 1 + r - s \ge q + 1$ since $r - s \ge 2$ else $\bar{a} \in N_2$.

Suppose $\bar{a}$ is of the form given in 2. Then for $s \ge 1$,

$$\left(1, \overset{q-r}{\ldots}, 1, 0.5, \overset{q-s}{\ldots}, 0.5, 0, \overset{r+s}{\ldots}, 0\right)$$
$$P$$
$$\left(0.5, \overset{q-r}{\ldots}, 0.5, 0, \overset{q-s}{\ldots}, 0, 1, \overset{r+s}{\ldots}, 1\right)$$
$$P$$
$$\left(0, \overset{q-r}{\ldots}, 0, 1, \overset{q-s}{\ldots}, 1, 0.5, \overset{r+s}{\ldots}, 0.5\right)$$

is a cycle since $q - r + q - s \geq q + 1 (r + s < q$ else $\bar{a} \in N_1), q - r + r + s \geq q + 1$, and $q - s + r + s \geq q + 1$. Suppose $s = 0$. Then

$$\left(1, \overset{q-r}{\ldots}, 1, 0.5, \overset{q}{\ldots}, 0.5, 0, \overset{r}{\ldots}, 0\right)$$

$$P$$

$$\left(0.5, \overset{q-r}{\ldots}, 0.5, 0, \overset{q-1}{\ldots}, 0, 0.5, 1, \overset{r}{\ldots}, 1\right)$$

$$P$$

$$\left(0, \overset{q-r}{\ldots}, 0, 1, \overset{q-1}{\ldots}, 1, 0, \right)$$

is a cycle since $2q - r \geq q + 1$, $q - r + r + 1 \geq q + 1$, and $q - 1 - r \geq q + 1$. ($s = 0$ implies $r \geq 2$ else $\bar{a} \in N_2$.)

5.4 Relation Spaces and Majority Rule

One of the implications of our argument in the previous section is that a fuzzy multi-dimensional public choice model can accommodate highly irregular shaped indifference curves, even those that are concave or multi-modal. Its ability to do relies on a homomorphism that permits a region of interest (a public choice model) to be mapped to a simpler region with a suitable and natural partial ordering where the results are determined and then faithfully transferred back to the original region of interest. We now present a formal proof of these results found in Mordeson and Clark.

Let R be a binary relation on a set X. The pair (X, R) is called a **relation space**. If $(\mathscr{A}, \tilde{R})$ and (X, R) are relation spaces, we give conditions when results from (X, R) can be faithfully carried back from (X, R) to $(\mathscr{A}, \tilde{R})$ by the preimage of a homomorphism of $(\mathscr{A}, R)$ onto (X, R). In particular, we give conditions involving the maximal sets and the Pareto sets of $(A, \tilde{R})$ and (X, R), Theorems 5.33, 5.35, and 5.36.

Definition 5.30 (*Homomorphism*) Let $(\mathscr{A}, \tilde{R})$ and (X, R) be relation spaces. Let f^* be a function of $\mathscr{A}$ into X. Then f^* is called a **homomorphism** of $(\mathscr{A}, \tilde{R})$ into (X, R) if $\forall a, b \in \mathscr{A}, (a, b) \in \tilde{R}$ if and only if $(f^*(a), f^*(b)) \in R$. If f^* maps $\mathscr{A}$ onto X, we say f^* maps $(\mathscr{A}, \tilde{R})$ onto (X, R). For all $(a, b) \in \tilde{R}$, we write

$$f^*((a, b)) = (f^*(a), f^*(b))$$

and

$$f^*(\tilde{R}) = \{(f^*((a, b)) \mid (a, b) \in \tilde{R}\}.$$

Let f^* be a **homomorphism** of $(\mathscr{A}, \widetilde{R})$ into (X, R). Then $\forall a, b \in \mathscr{A}, (a, b) \in \widetilde{R}$ if and only if $(f^*(a), f^*(b)) \in R$. Thus if $a, a', b, b' \in \mathscr{A}$ and $f^*(a) = f^*(a'), f^*(b) = f^*(b')$, it is not possible that $(a, b) \in \widetilde{R}$ and $(a', b') \notin \widetilde{R}$.

Proposition 5.31 *Let f^* be a homomorphism of $(\mathscr{A}, \widetilde{R})$ onto (X, R). Then $f^*(\widetilde{R}) = R$.*

Proof Clearly, $f^*(\widetilde{R}) \subseteq R$. Let $(x, y) \in R$. Since f^* maps $\mathscr{A}$ onto X, there exists $a, b \in \mathscr{A}$ such that $f^*(a) = x$ and $f^*(b) = y$. Thus,

$$(x, y) = (f^*(a), f^*(b)) = f^*((a, b)) \in f^*(\widetilde{R}).$$

Proposition 5.32 *Let f^* be a homomorphism of $(\mathscr{A}, \widetilde{R})$ onto (X, R). Then $\forall a, b \in \mathscr{A}, (a, b) \in \widetilde{P}$ if and only if $(f^*(a), f^*(b)) \in P$.*

Proof Let $a, b \in \mathscr{A}$. Then

$$
\begin{aligned}
(a, b) \in \widetilde{P} &\Leftrightarrow (a, b) \in \widetilde{R} \text{ and } (b, a) \notin \widetilde{R} \\
&\Leftrightarrow (f^*(a), f^*(b) \in R \text{ and } (f^*(a), f^*(b)) \notin R \\
&\Leftrightarrow (f^*(a), f^*(b)) \in P .
\end{aligned}
$$

Theorem 5.33 *Let f^* be a homomorphism of $(\mathscr{A}, \widetilde{R})$ onto (X, R). Then $f^*(M(\widetilde{R}, \mathscr{A})) = M(R, X)$. Furthermore, $f^{*-1}(M(R, X)) = M(\widetilde{R}, \mathscr{A})$.*

Proof We have that

$$
\begin{aligned}
a \in M(\widetilde{R}, \mathscr{A}) &\Leftrightarrow \forall b \in \mathscr{A}, \ a\widetilde{R}b \Leftrightarrow \Leftrightarrow \\
&\Leftrightarrow \forall f^*(b) \in X, \ f^*(a)Rf^*(b) \\
&\Leftrightarrow f^*(a) \in M(R, X),
\end{aligned}
$$

where the latter equivalence holds since f^* maps $\mathscr{A}$ onto X. Thus if $f^*(a) \in f^*(M(\widetilde{R}, \mathscr{A}))$, then $a \in M(\widetilde{R}, \mathscr{A})$. Hence $f^*(a) \in M(R, X)$. Thus

$$f^*(M(\widetilde{R}, \mathscr{A})) \subseteq M(R, X).$$

Let $x \in M(R, X)$. Then $\forall y \in X, xRy$. Let $a \in \mathscr{A}$ be such that $f^*(a) = x$. Let $b \in \mathscr{A}$. Then $f^*(a)Rf^*(b)$ since $x = f^*(a)$ and $x \in M(R, X)$. Hence $a\widetilde{R}b$ by Definition 5.30 Thus $a \in M(\widetilde{R}, \mathscr{A})$ and so $x = f^*(a) \in f^*(M(\widetilde{R}, \mathscr{A}))$. Hence

$$M(R, X) \subseteq f^*(M(\widetilde{R}, \mathscr{A})).$$

Clearly, $f^{*-1}(M(R, X)) \supseteq M(\widetilde{R}, \mathscr{A})$. Let $a \in f^{*-1}(M(R, X))$. Suppose $\exists b \in \mathscr{A}$ such that $(a, b) \notin \widetilde{R}$. Then $(f^*(a), f^*(b)) \notin R$ since f^* is a homomorphism. Thus $f^*(a) \notin M(R, X)$ a contradiction of $a \in f^{*-1}(M(R, X))$. Hence $(a, b) \in \widetilde{R} \ \forall b \in \mathscr{A}$. Thus $a \in M(\widetilde{R}, \mathscr{A})$. Hence

$$f^{*-1}(M(R,X)) \subseteq M(\widetilde{R}, \mathscr{A}).$$

Let $(\mathscr{A}, \widetilde{R}_i)$ be a relation space, $i = 1, \ldots, n$. Let f_i^* be a homomorphism of $(\mathscr{A}, \widetilde{R}_i)$ onto (X, R_i), $i = 1, \ldots, n$. Then $R_i = f_i^*(\widetilde{R}_i)$, $i = 1, \ldots, n$ by Proposition 5.31.

Definition 5.34 Let $\widetilde{f}$ be an aggregation rule on $(\mathscr{A}, (\widetilde{R}_1, \ldots, \widetilde{R}_n))$ and let f be an aggregation rule on $(X, (R_1, \ldots, R_n))$. Let f_i^* be a homomorphism of $(\mathscr{A}, \widetilde{R}_i)$ onto (X, R_i), $i = 1, \ldots, n$. Let f^* be a homomorphism of $(\mathscr{A}, \widetilde{f}((\widetilde{R}_1, \ldots, \widetilde{R}_n)))$ onto $(X, f((R_1, \ldots, R_n)))$. Then f^* is said to **preserve the pair** $(\widetilde{f}, f)$ **with respect to** $(f_1^*, \ldots, f_n^*)$ if $f^*(\widetilde{f}(\widetilde{R}_1, \ldots, \widetilde{R}_n)) = f((R_1, \ldots, R_n))$.

Theorem 5.35 *Let $\widetilde{f}$ be an aggregation rule on $(\mathscr{A}, (\widetilde{R}_1, \ldots, \widetilde{R}_n))$ and let f be an aggregation rule on $(X, (R_1, \ldots, R_n))$. Let f_i^* be a homomorphism of $(\mathscr{A}, \widetilde{R}_i)$ onto (X, R_i), $i = 1, \ldots, n$. Let f^* be a homomorphism of $(\mathscr{A}, \widetilde{f}((\widetilde{R}_1, \ldots, \widetilde{R}_n)))$ onto $(X, f((R_1, \ldots, R_n)))$ such that f^* preserves $(\widetilde{f}, f)$ w.r.t. $(f_1^*, \ldots, f_n^*)$. Then $f^*(PS_N(\widetilde{R})) = PS_N(R)$, where $\widetilde{R} = \widetilde{f}((\widetilde{R}_1, \ldots, \widetilde{R}_n))$ and $R = f((R_1, \ldots, R_n))$. Furthermore, $f^{*-1}(PS_N(\widetilde{R})) = PS_N(R)$.*

Proof We have that

$$a \in PS_N(\widetilde{R}) \Leftrightarrow \forall b \in \mathscr{A}, \left(\exists i \in N, \ b\widetilde{P}a \Rightarrow \exists j \in N, \ a\widetilde{P}b \right).$$

Thus if $f^*(a) \in f^*(PS_N(\widetilde{R}))$, then $a \in PS_N(\widetilde{R})$. Hence $f^*(a) \in PS_N(R)$. Thus

$$f^*(PS_N(\widetilde{R})) \subseteq PS_N(R).$$

Let $x \in PS_N(R)$. Let $y \in X$. If $\exists i \in N$ such that yP_ix, then $\forall j \in N$ such that xR_jy. Let $a \in \mathscr{A}$ be such that $f^*(a) = x$. Let $b \in \mathscr{A}$. Then $f^*(b)P_if^*(a) \Leftrightarrow b\widetilde{P}_ia$ and $f^*(a)P_jf^*(b) \Leftrightarrow a\widetilde{P}_jb$. Hence if $\exists i \in N$ such that $b\widetilde{P}_ia$, then $\exists j \in N$ such that $a\widetilde{P}_jb$. Thus $a \in PS_N(\widetilde{R})$ and so $x = f^*(a) \in f^*(PS_N(\widetilde{R}))$. Hence

$$PS_N(R) \subseteq f^*(PS_N(\widetilde{R})).$$

Clearly, $f^{*-1}(PS_N(R)) \supseteq PS_N(\widetilde{R})$. Let $a \in f^{*-1}(PS_N(R))$. Suppose $a \notin PS_N(\widetilde{R})$. Then it's not the case that $\forall b \in A \ \exists i \in N, \ b\widetilde{R}_ia \Rightarrow \exists j \in N, \ a\widetilde{P}_jb$. Thus $\exists b \in a$ such that it is not the case that

$$\exists i \in N \ b\widetilde{R}_ia \Rightarrow \exists j \in N \ a\widetilde{P}_jb.$$

Thus if $\exists i \in N$ such that $b\widetilde{P}_ia$, then it is not the case that $\exists j \in N$ such that $a\widetilde{P}_jb$ and so $b\widetilde{R}_ja \ \forall j \in N$. Hence $(b, a) \in \widetilde{R}_i \ \forall i \in N$ and so $(f^*(a), f^*(b)) \in R_i\forall \in N$. Thus $(b, a) \in \widetilde{R}_i \ \forall i \in N$ and so $(f^*(b), f^*(a)) \in R_i\forall i \in N$. Hence $f^*)(a) \notin PS_N(R)$. However, this contradicts the fact that $a \in f^{*-1}(PS_N(R))$. Thus $a \in PS_N(\widetilde{R})$ and so $f^{*-1}(PS_N(\widetilde{R})) \subseteq PS_N(R)$.

Theorem 5.36 *Let $\tilde{f}$ be an aggregation rule on $(\mathscr{A}, (\tilde{R}_1, \ldots, \tilde{R}_n))$ and let f be an aggregation rule on $(X, (R_1, \ldots, R_n))$. Let f_i^* be a homomorphism of $(\mathscr{A}, \tilde{R}_i)$ onto (X, R_i), $i = 1, \ldots, n$. Let f^* be a homomorphism of $(\mathscr{A}, \tilde{f}((\tilde{R}_1, \ldots, \tilde{R}_n)))$ onto $(X, f((R_1, \ldots, R_n)))$ such that f^* preserves $(\tilde{f}, f)$ w.r.t. $(f_1^*, \ldots, f_n^*)$. Then $\tilde{f}$ is a simple majority rule if and only if f is a simple majority rule.*

Proof By Proposition 5.32, $\forall a, b \in \mathscr{A}$, $(a, b) \in \tilde{P}_i$ if and only if $(f^*(a), f^*(b)) \in P_i$, $i = 1, \ldots, n$. Thus it follows that

$$\left| \tilde{P}\left(a, b; \tilde{f}((\tilde{R}_1, \ldots, \tilde{R}_n))\right) \right| = \left| P\left(f^*(a), f^*(b)); f((R_1, \ldots, R_n))\right) \right|.$$

The desired result is now immediate.

5.5 A Consideration of Different Definitions of Fuzzy Covering Relations

There are several definitions of the covering set. When players' preferences are strict, these definitions all result in the same uncovered set. However, that is not the case when indifference is permitted (Penn 2006). The fuzzy discrete preference that we introduced earlier in this book permits the representation of thick indifference in players' preferences in multi-dimensional public choice models. In this concluding section, we consider the effect of thick indifference on several of the most frequently used definitions of covering relations found in the literature.

We assume that each player's preferences is a fuzzy subset of the original region of interest $\mathbb{R}^{2+}$, the first quadrant of the plane, and take on the values 0 and 1 and possibly $0.25, 0.5, 0.75$. The plane is mapped to the lattice T^n, where $T = \{0, 0.25, 0.5, 0.75, 1\}$. We begin by showing the relationship between the partial order in T^n and various covering relations, e.g., Theorem 5.97. We show in Theorem 5.40 that for any subset X of U, there exists a unique minimal generating set S_X for X and that S_X has the property that for all $y \in S_X$, there does not exist $x \in S_X$ such that $x \prec y$.

Our examination of definitions of the covering relationship begins with the most common definition encountered in the literature. We label this covering relation C. C yields an uncovered set, $UC(R)$, for which there exists $x, y \in UC(R)$ such that $x \prec y$. Consequently, x is not Pareto efficient, i.e., $x \in UC(R)$, but $x \notin PS_N(R)$, the Pareto set of R. It is not reasonable to expect that a majority of players will settle for a Pareto deficient outcome.

We then consider an alternative definition of the covering relation developed in Mordeson et al. (2011), C_4, which permits x to be weakly preferred to y. C requires that x be strictly preferred to y. Our main result, Theorem 5.55, demonstrates that under a mild condition, $UC_4(R) = S_{UC_4(R)}$. Hence under this condition, we have for all $y \in UC_4(R)$ that there does not exist $x \in UC_4(R)$ such that $x \prec y$. We also show the relationship between C and C_4 in Corollary 5.56.

We next examine three other definitions of the covering relation: the Fishburn set, C_2; a combination of C_2 and C_4, C_7, in Mordeson et al. (2011), and a definition permitting x to cover y and y to cover x. We demonstrate that none of the three produce the desired result.

We conclude by demonstrating that C_4 is subject to the "vulnerability to holes" (Mordeson et al. 2011). When immediate predecessors of elements of the Pareto set are not in the set of alternatives, non-Pareto efficient alternatives may result in the uncovered set.

5.5.1 Covering Relations and Minimal Generating Sets

We continue to assume that X is a subset of a universe U of interest. Let $\mathscr{R}$ denote the set of all binary relations on X which are reflexive, complete and transitive.

Miller (1980) contains the most oft used definitions of the covering relation. We give that definition here. We henceforth refer to this definition of the covering relation as C.

Definition 5.37 Let R be a binary relation on X. Define the binary relation C on X by $\forall x, y \in X$, xCy if and only if xPy and $\forall z \in X$, yPz implies xPz. If xCy, then x is said to C-cover y (or simply, x is said to cover y). If no such x exists, then y is said to be C-uncovered (or simply uncovered). Let $UC(R)$ denote the set of all uncovered elements of X with respect to C and R.

Proposition 5.38 *Let $x, c \in X$. If x is uncovered and $x \in \langle c \rangle$, then c is uncovered.*

Proof Suppose c is covered. Then $\exists d \in X$ such that dPc and $\forall z \in X$, cPz implies dPz. Since $x \in \langle c \rangle$, dPx. Suppose xPz for some $z \in X$. Then cPz since $x \preceq c$. Since d covers c, dPz. Thus d covers x.

We show in Example 5.46 that it is possible for $UC(R)$ to contain elements x for which there exists y in $UC(R)$ such that $x \prec y$. This motivates the development of the following ideas.

Definition 5.39 Let $S \subseteq X$. Then S is called a **generating set** for X if $\langle S \rangle \supseteq X$. S is called a **minimal generating set** for X if S is a generating set for X and $\forall s \in S$, $\langle S \backslash \{s\} \rangle \subset \langle S \rangle$.

Theorem 5.40 *X has a unique minimal generating set.*

Proof Let $S = \{s \in X \mid \nexists c \in X, s \prec c\}$. ($S$ exists since X is finite.) Thus $\forall x \in X$, $\forall s \in S$, not $s \prec x$. Hence $\forall x \in X$, either $\exists s \in S$ such that $x \preceq s$ and so $x \in \langle s \rangle$ or $\bar{x}$ and every $s \in S$ are incomparable. In the latter case, $\nexists s \in S$ such that $x \preceq s$ and so $\exists y \in X$ such that $x \preceq y$ and $\nexists z \in X$ such that $y \prec z$. Thus $y \in S$, a contradiction. Hence since $x \in \langle \{s\} \rangle$, $X \subseteq \cup_{s \in S} \langle \{s\} \rangle = \langle S \rangle$. Thus S is a generating set for X. Suppose $\exists s \in S$ such that $\langle S \backslash \{s\} \rangle = \langle S \rangle$. Then $s \in \langle S \backslash \{s\} \rangle = \cup_{r \in S \backslash \{s\}} \langle \{r\} \rangle$. Hence $s \in \langle \{r\} \rangle$ for some $r \in S \backslash \{s\}$. Thus $s \preceq r$. However, since $\nexists t \in X$ such that $s \prec t$, $s = r$ and so $s \in S \backslash \{s\}$, a contradiction. Hence no such s exists so S is a minimal generating set for X. Suppose S' is also a minimal generating set for X.

Then $\langle S \rangle = \langle X \rangle = \langle S' \rangle$. Thus $\forall s' \in S', s' \in \langle S \rangle$ and so $\exists s \in S$ such that $s' \in \langle \{s\} \rangle$, i.e., $s' \preceq s$. Since $\nexists x \in X$ such that $s' \prec x, s' = s$. Hence $s' \in S$ and so $S' \subseteq S$. Similarly, $S \subseteq S'$. Thus $S' = S$. Hence S is unique.

Let $UC(R)$ be the uncovered set of X with respect to R. Let S be the unique minimal generating set of $UC(R)$. Then $\forall x \in UC(R) \backslash S, x \prec s$ for some $s \in S$. We show in Example 5.46 that it is possible for $UC(R)$ to strictly contain its unique minimal generating set.

Definition 5.41 Let $s \in X$. Then s is said to be **rationally uncovered** if $s \in UC(R)$ and $\forall x \in UC(R), s \preceq x$ implies $s = x$.

Proposition 5.42 *Let S be a subset of X. Then S is the set of all rationally uncovered elements of X if and only if S is the unique minimal generating set of $UC(R)$.*

Let $X \subseteq T^n$. Let R_i be the binary relation on X defined by $\forall s = (s_1, \ldots, s_n), x = (x_1, \ldots, x_n) \in X$, $sR_i x$ if and only if $s_i \geq x_i$ for $i = 1, \ldots, n$. Let $R(x, s; \rho) = \{i \in N \mid xR_i s\}$, where $\rho = (R_1, \ldots, R_n)$ and $N = \{1, \ldots, n\}$.

Proposition 5.43 *Suppose $X \subseteq T^n$. Let $s \in X$. Then s is rationally uncovered if and only if $s \in UC(R)$ and $\forall x \in UC(R), R(x, s; \rho) = N$ implies $R(s, x; \rho) = N$.*

Proof $s \preceq x \Leftrightarrow s_i \leq x_i, i = 1, \ldots, n \Leftrightarrow xR_i s, i = 1, \ldots, n \Leftrightarrow R(x, s; \rho) = N$. Thus $s = x \Leftrightarrow R(x, s; \rho) = N = R(s, x; \rho)$.

Definition 5.44 (*Choice Function*) Let $C : \mathscr{P}^*(X) \to \mathscr{P}^*(X)$. Then C is called a **choice function** on X if $\forall S \in \mathscr{P}^*(X), C(S) \subseteq S$.

We next introduce a binary relation on X induced by a choice function on X. This leads us to Example 5.46 where, we show that there exists $x, y \in UC(R)$ such that $x < y$.

Definition 5.45 Let C be a choice function on X. Let $R = \{(x, y) \in X \times X \mid x \in C(S), y \in S$ for some $S \in \mathscr{P}^*(X)\}$. Then R is called the **binary operation induced** by C.

Since $C(\{x\}) = \{x\}$ for all $x \in X, R$ is reflexive. Since $\emptyset \neq C(\{x, y\}) \subseteq \{x, y\}, R$ is complete.

Let C_{mgs} be the choice function on X defined by for all nonempty subsets Y of X, $C_{mgs}(Y) = S_Y$, where S_Y is the unique minimal generating set of Y. In the next example, we show that it is possible for $UC(R) \supset S_{UC(R)}$.

Example 5.46 Let $U = T^3$. Let $X \subseteq U$ and R be a binary relation on X such that

$$PS_N(R) = \{(1, 0.75, 0.5), (0.75, 0.5, 1), (0.5, 1, 0.75)\}$$

and $X = \langle PS_N(R) \rangle$. Let $UC(R)$ denote the uncovered set of X with respect to R. Then we show that

$$C_{mgs}(UC(R)) = \{(1, 0.75, 0.5), (0.75, 0.5, 1), (0.5, 1, 0.75)\}.$$

It suffices to show that $(1, 0.75, 0.5)$, $(0.75, 0.5, 1)$, $(0.5, 1, 0.75)$ are uncovered since $\nexists \bar{x} \in X$ such that $\bar{x} \succ$ any of these elements. Suppose $(a, b, c)P(1, 0.75, 0.5)$. Then $b = 1$ and $c > 0.5$. Let $\bar{z} = (0.75, 0.5, 1)$. Then $(1, 0.75, 0.5)P\bar{z}$. However not $(a, b, c)P\bar{z}$ else $a = 1$, but then $(a, b, c) \notin X$. Thus $(1, 0.75, 0.5)$ is uncovered. Similarly, $(0.75, 0.5, 1)$ and $(0.5, 1, 0.75)$ are uncovered. We next show that $(0.75, 0.75, 0.5) \in UC(R)$ and so $UC(R) \supset S_{UC(R)}$. Suppose $\bar{x}C(0.75, 0.75, 0.5)$. Then $\bar{x}P(0.75, 0.75, 0.5)$. Thus $x_1 = 1, x_2 = 1$ or $x_1 = 1, x_3 \geq 0.75$ or $x_2 = 1, x_3 \geq 0.75$, where $\bar{x} = (x_1, x_2, x_3)$. The first two cases are impossible. However $(a, 1, 0.75)P(0.75, 0.75, 0.5)$, where $a \leq 0.5$ (since $(a, 1, 0.5) \notin X$ if $a > 0.5$). Now $(0.75, 0.75, 0.5)P\bar{z}$, but not $(a, 1, 0.75)P\bar{z}$, where $\bar{z} = (0.5, 0.5, 1)$. Thus $(a, 1, 0.75)$ does not cover $(0.75, 0.75, 0.5)$. Hence $(0.75, 0.75, 0.5) \in UC(R)$.

Let R_{mgs} denote the relation induced by C_{mgs}. Let

$$Y = \{(0.25, 0, 0)\} \cup \{(0, b, c) \mid (0, b, c) \in X\}.$$

Then

$$C_{mgs}(Y) = \{(0.25, 0, 0), (0, 1, 0.75), (0, 0.5, 1)\}.$$

Thus,

$$\{(0.25, 0, 0), (0, 1, 0.75), (0, 0.5, 1)\} \times Y \subseteq R_{mgs}.$$

5.5.2 Covering Relation C_4 with $n \geq 3$

Since there is no reason to believe that a majority would choose a Pareto deficient alternative, the fact that $UC(R)$ contains elements that are not Pareto efficient is problematic. In what follows, we examine alternative definitions of covering. In the remainder of the section, we let $U = T^n$.

We begin with a definition of the covering relation, C_4, that results in an uncovered set referred to in the literature as the F_D set (Austen-Smith and Banks 1999).

Definition 5.47 Mordeson et al. (2011) Defines the binary relation C_4 on X by $\forall x, y \in X, xC_4y \Leftrightarrow xRy$ and $\forall z \in X[yPz \Rightarrow xPz]$ and $\exists w \in X[xPw$ and not $yPw]$.

The primary difference between the definition of the covering relation C_4 and the conventional definition of covering, C, is that x is no longer strictly preferred to y. It may be weakly preferred. This is a necessary modification given preference indifference. As we will demonstrate, the modification results in an uncovered set that does not contain Pareto deficient elements.

Let $X \subseteq T^n$ and let $J_{n/2} = \{x = (x_1, \ldots, x_n) \in T^n \mid x_i \in \{0, 1\}, i = 1, \ldots, n, |\{i \in N \mid x_i = 0\}| \geq n/2$ and $\exists j \in N$ such that $x_j = 1\}$.

Proposition 5.48 (Mordeson et al. 2011) *Let $x, y \in X$. If xC_4y, then either $x \succ y$ or $y \in \langle J_{n/2} \rangle$.*

Proof Suppose $x \preceq y$. Then clearly $\nexists w \in X$ such that xPw and not yPw, a contradiction since xC_4y. Thus either $x \succ y$ or x and y are not comparable with respect to $\preceq$. Suppose x and y are not comparable with respect to $\preceq$, where $x = (x_1, \ldots, x_n)$ and $(y_1, \ldots, y_n)$. Then $\exists i, j \in N$ such that $x_i > y_i$ and $x_j < y_j$. Since xC_4y, xRy. Thus not yPx. Hence strictly fewer than $[[n/2]] + 1$ of the y_i are strictly greater than the corresponding x_i. There is no loss in generality in assuming $y_1 \leq x_1, \ldots, y_r \leq x_r$ and $y_{r+1} > x_{r+1}, \ldots, y_n > x_n$, where $n - 2 < [[n/2]] + 1$. Suppose $y_1 = \cdots = y_{s-1} = 0$ for $s \geq 1$. (The case $s = 1$ says no $y_i = 0$.) We show $s - 1 \geq n/2$. Assume $s - 1 < n/2$. Then $n - s + 1 > n/2$. Let t be such that $r - t - s + 1 + n - r = [[n/2]] + 1$. Now let $y'_s, \ldots, y'_{r-t}$ be such that $y_s > y'_s \geq 0, \ldots, y_{r-t} > y'_{r-t} \geq 0$. Let

$$z = (x_1, \ldots, x_{s-t}, y'_s, \ldots, y'_{r-t}, x_{r-t+1}, \ldots, x_r, x_{r+1}, \ldots, x_n)$$

Now $n - t - s + 1 = [[n/2]] + 1$. Thus

$$t + s - 1 = \begin{cases} [[n/2]] - 1 & \text{if } n \text{ is even,} \\ [[n/2]] & \text{if } n \text{ is odd.} \end{cases}$$

Since $y'_s < y_s \leq x_s, \ldots, y'_{r-t} < y_{r-t} \leq x_{r-t}$ and $X = \langle X \rangle$, $z \in X$. Now $t + s - 1$ is the number of components x_i is strictly greater than the corresponding components of z. Thus not xPz. However, $r - t - s + 1 + n - r = [[n/2]] + 1$ (see above) is the number of components of y that are strictly greater than the corresponding components of z. Thus yPz, contradicting the hypothesis that xC_4y. Thus $s - 1 \geq n/2$. Hence $y \in \langle J_{n/2} \rangle$.

Proposition 5.49 (Mordeson et al. 2011) *Let $x, y \in X$. If $x \succ y$, then either ($\exists w \in X$ such that xPw and not yPw) or $x \in \langle J_{n/2} \rangle$.*

Proof Suppose x and y differ in $[[n/2]] + 1$ or more components. Then let $w = y$. Suppose x and y differ in fewer than $[[n/2]] + 1$ components. There is no loss in generality in assuming that $x = (x_1, \ldots, x_n)$ and $y = (y_1, \ldots, y_n)$, where $x_1 = y_1, \ldots, x_r = y_r, x_{r+1} > y_{r+1}, \ldots, x_n > y_n$, and $n - r \leq [[n/2]]$. Thus

$$r \geq n - [[n/2]] = \begin{cases} [[n/2]] & \text{if } n \text{ is even,} \\ [[n/2]] + 1 & \text{if } n \text{ is odd.} \end{cases}$$

Suppose $x \notin \langle J_{n/2} \rangle$. Then fewer than $n/2$ of the $x_i = 0$, say $x_1 = \cdots = x_s = 0$, where $s < n/2$. Thus $s < r$. Let t be such that $t - s + n - r = [[n/2]] + 1$. Let

$$w = (x_1, \ldots, x_s w_{s+1}, \ldots, w_t, y_{t+1}, \ldots, y_r, y_{r+1}, \ldots, y_n),$$

where $x_{s+1} > w_{s+1} \geq 0, \ldots, x_t > w_t \geq 0$. Since $x_{s+1} > w_{s+1}, \ldots, x_t > w_t$ and $x_i \geq y_i, \ldots, x_n \geq y_n$ and $X = \langle X \rangle$, $w \in X$. Now $t - s + n - r$ of the components of x are strictly greater than the corresponding components of w. Thus xPw. Now $t - s$ components of y are greater than the corresponding components of w and $t - s \leq [[n/2]]$. Hence not yPw.

Proposition 5.50 (Mordeson et al. 2011) *Let $x \in X$. Suppose x is C_4-uncovered. Then either $x \in PS_N(R)$ or $x \in \langle J_{n/2} \rangle$.*

Proof Suppose $x \notin PS_N(R)$ and $x \notin \langle J_{n/2} \rangle$. Since $x \notin PS_N(R)$, there exists $y \in X$ with $y \succ x$ by (2) of Corollary 5.8. (Recall $PS_N(R) = PS_N(\overline{\rho})$ there.) By Proposition 5.49, either xC_4y or $x \in \langle J_{n/2} \rangle$. But since $x \notin \langle J_{n/2} \rangle$ and $y \succ x$, we have $y \notin \langle J_{n/2} \rangle$. On the other hand, yC_4x contradicts that x is C_4-uncovered.

In words, all elements in the uncovered set are either in the Pareto set or they are in a special set of alternatives, $\langle J_{n/2} \rangle$. The fact that a special set of alternatives is possible would appear to severely limit the likelihood that covering relation C_4 will produce a reasonably constructed uncovered set. However, $J_{n/2}$ comprises a set of relatively trivial alternatives. Each alternative in the set is considered by less than half of the players to be perfectly in the set of ideal points ($\alpha = 1$). Moreover, all remaining players consider the alternative to be entirely not in the set of ideal points ($\alpha = 0$). Thus, none of the alternatives in either $J_{n/2}$ or $\langle J_{n/2} \rangle$, can defeat any alternative by majority vote. Hence they cannot cover any other alternative. Therefore, if $X \subseteq \langle J_{n/2} \rangle$, then every element of X is C-uncovered.

We further note that if there is even one alternative for which a majority of players express the slightest degree of preference (that is, a majority prefer the alternative at $\alpha > 0$), then every alternative in $\langle J_{n/2} \rangle$ is covered by that alternative and therefore cannot be in the uncovered set. Even if the alternative ties elements in $\langle J_{n/2} \rangle$, it can defeat all alternatives lying in the region outside of the support for all players' preferences. Hence it covers elements in $\langle J_{n/2} \rangle$. Thus elements in $\langle J_{n/2} \rangle$ are uncovered if and only if $\langle J_{n/2} \rangle$ are the only elements in X. In effect, no alternatives are supported by a majority at any α-level.

The following statement summarizes the foregoing. If x is C_4-uncovered and $x \in \langle J_{n/2} \rangle$, then every element of X is uncovered.

Proposition 5.51 $S_{UC(R)} \subseteq UC_4(R) \cup J_{n/2}$.

Proof Let $y \in S_{UC(R)}$. If $\exists x \in X$ such that xC_4y, then $x \succ y$ or $y \in \langle J_{n/2} \rangle$ by Proposition 5.48. Since $y \in S_{UC(R)}$, $\nexists x \in X$ such that $x \succ y$. Hence $y \in \langle J_{n/2} \rangle$. Since $\nexists x \in X$ such that $x \succ y$, $y \in J_{n/2}$.

Proposition 5.52 $UC_4(R) \subseteq S_{UC_4(R)} \cup \langle J_{n/2} \rangle$.

Proof Let $y \in UC_4(R)$. Then $\exists x \in S_{UC_4(R)}$ such that $y \in \langle \{x\} \rangle$. Hence $y = x$ or $y \prec x$. If $y = x$, then of course $y \in S_{UC_4(R)}$. Suppose $y \prec x$. Then xRy (or possibly even xPy) and clearly $\forall z \in X$, $yPz \Rightarrow xPz$. By Proposition 5.49, either $\exists w \in X[xPw$ and not $yPx]$ or $x \in \langle J_{n/2} \rangle$. If such a w exists, then xC_4y, contradicting $y \in UC_4(R)$. Thus $x \in \langle J_{n/2} \rangle$ and so $y \in \langle J_{n/2} \rangle$.

Theorem 5.53 $S_{UC(R)} \cup \langle J_{n/2} \rangle = S_{UC_4(R)} \cup \langle J_{n/2} \rangle$.

Proof By Propositions 5.51 and 5.52, $S_{UC(R)} \subseteq UC_4(R) \cup J_{n/2} \subseteq S_{UC_4(R)} \cup \langle J_{n/2} \rangle$. Thus it follows that $S_{UC(R)} \subseteq S_{UC_4(R)} \cup J_{n/2}$. Let $y \in S_{UC_4(R)}$. Suppose $y \notin S_{UC(R)}$. Then $y \notin \langle S_{UC(R)} \rangle$ since $\nexists x \in X$ such that $x \succ y$. Hence y is C-covered. Thus $\exists x \in X$ such that xPy and $\forall z \in X, yPz \Rightarrow xPz$, i.e., xCy. Since not $x \succ y$ and xPy, at least one of the components of x is $<$ the corresponding component of y and at least $[[n/2]] + 1$ of the components of x are $>$ the corresponding components of y, say $x = (x_1, \ldots, x_{r-1}, x_r, x_{r+1}, \ldots, x_n)$ and $y = (y_1, \ldots, y_{r-1}y_r, y_{r+1}, \ldots, y_n)$ with $x_i \leq y_i$ for $i = 1, \ldots, r-1$ and $x_r < y_r$, and $x_i > y_i$ for $i = r+1, \ldots, n$. Then $n - r \geq [[n/2]] + 1$. Suppose $[[n/2]]$ of $y_{r+1}, \ldots, y_n$ are strictly greater than 0, say $y_{r+1}, \ldots, y_s$. Let $z = (x_1, \ldots, x_{r-1}, x_r, 0, \ldots, 0, x_{s+1}, \ldots, x_n)$. Then yPz since at least $[[n/2]]+1$ of the components of y are greater than the corresponding components of z. However not xPz since fewer that $[[n/2]]$ of the components of x are strictly greater than the corresponding components of z. Thus not xCy, a contradiction. Hence $n - [[n/2]] + 1$ of the components of y equal 0. Thus $y \in \langle J_{n/2} \rangle$. Hence $S_{UC_4(R)} \subseteq S_{UC(R)} \cup \langle J_{n/2} \rangle$. Thus $S_{UC(R)} \cup \langle J_{n/2} \rangle = S_{UC_4(R)} \cup \langle J_{n/2} \rangle$.

Proposition 5.54 *Either* $UC_4(R) \cap \langle J_{n/2} \rangle = \emptyset$ *or* $UC_4(R) = X \subseteq \langle J_{n/2} \rangle$.

Proof Let $x \in \langle J_{n/2} \rangle$. Suppose $X \backslash \langle J_{n/2} \rangle \neq \emptyset$. Let $y \in X \backslash \langle J_{n/2} \rangle$. Then yRx and $\forall z \in X, xPz \Rightarrow yPz$ vacuously since $\nexists z \in X$ such that xPz. Now $\exists w \in X$ such that yPw and not xPw, namely $w = \bar{0}$. Thus yC_4x. Hence $x \notin UC_4(R)$. Suppose $X \backslash \langle J_{n/2} \rangle = \emptyset$. Then $X \subseteq \langle J_{n/2} \rangle$. Let $x, y \in \langle J_{n/2} \rangle$. Then $\nexists w \in X$ such that xPw and not yPw since $\nexists w \in X$ such that xPw due to the fact that the number of nonzero components of x is less than or equal to $[[n/2]]$. Thus not xC_4y. Hence $UC_4(R) = X$.

Theorem 5.55 *Suppose* $X \nsubseteq \langle J_{n/2} \rangle$. *Then* $UC_4(R) = S_{UC_4(R)}$.

Proof The proof follows by Propositions 5.52 and 5.54. (The result also follows from Propositions 5.50 and 5.54.)

Corollary 5.56 *Suppose* $X \nsubseteq \langle J_{n/2} \rangle$. *Then* $S_{UC(R)} \cup J_{n/2} = UC_4(R) \cup J_{n/2}$.

Proof $UC_4(R) \subseteq S_{UC(R)} \subseteq UC_4(R) \cup J_{n/2}$, where the former inclusion follows from Theorem 5.53 and Proposition 5.54 and the latter inclusion follows Proposition 5.51. Thus $UC_4(R) \cup J_{n/2} \subseteq S_{UC(R)} \cup J_{n/2} \subseteq UC_4(R) \cup J_{n/2}$ from which the desired result is immediate.

By Corollary 5.56 and Proposition 5.54, $UC_4(R) \subseteq S_{UC(R)}$.

Corollary 5.57 *Suppose* $X \nsubseteq \langle J_{n/2} \rangle$. *Then* $UC(R) \backslash \langle J_{n/2} \rangle = \langle UC_4(R) \rangle$.

Proof Since

$$\langle S_{UC(R)} \backslash J_{n/2} \rangle = \bigcup_{s \in S_{UC(R)} \backslash J_{n/2}} \langle \{s\} \rangle$$

$$= \bigcup_{s \in S_{UC(R)}} \langle \{s\} \rangle \backslash \bigcup_{s \in J_{n/2}} \langle \{s\} \rangle$$

$$= \langle S_{UC(R)} \rangle \backslash \langle J_{n/2} \rangle,$$

the desired result follows from Proposition 5.54 and Corollary 5.56.

Suppose U^*C_4 is the C_4-uncovered set for $\langle X \rangle$. Suppose that X has holes, i.e., $X \subset \langle X \rangle$. Note that $U^*C(R) \subseteq X$ since $U^*C(R) \subseteq PS_N(R)$. Suppose $w \in \langle X \rangle \backslash X$ and that for $x \in U^*C_2(R)$, xPw and not yPw for some $y \in X$, but no other w exists. Then x is not in $UC_4(R)$, the C_4-uncovered set for X. However $x \in U^*C(R)$ the uncovered set for X. We illustrate these ideas in the following example.

Example 5.58 Let

$$PS_N(R) = \{(1, 0, 0), (0, 1, 0), (0.25, 0, 1)\}$$

and also let

$$X = \langle PS_N(R) \rangle \backslash \{(0, 0, 0.75)\}.$$

That is,

$$\widetilde{A}_1^0 \cap \widetilde{A}_2^0 \cap \widetilde{A}_3^{0.75} = \widetilde{A}_1^0 \cap \widetilde{A}_2^0 \cap \widetilde{A}_3^1$$

or

$$\widetilde{A}_3^{0.75} = \widetilde{A}_3^1.$$

Let $x = (0.25, 0, 1)$ and $y = (0.25, 0, 0.75)$. Then $\forall z \in X$, yPz implies xPz. However, $\nexists w \in X$ such that xPz and not yPz since $(0, 0, 0.75) \notin X$. However $(0, 0, 0.75) \in \langle PS_N(R) \rangle$.

5.5.3 Covering Relation C_2

Another definition of the covering relation, which we label C_2, is offered by the Fishburn function or Fishburn set (Bordes 1983; Fishburn 1977; Richelson 1980, 1981). C_2 has been widely used by scholars Shepsle and Weingast (1987), Cox (1987), Bianco et al. (2004), Epstein (1988). In what follows, we modify C_2 by permitting x to be weakly preferred to y.

5.5.3.1 Covering Relation C_2 Modified to Permit Weak Preference

Define the binary relation C_2 on X by $\forall x, y \in X$, xC_2y if and only if xRy, $\forall z \in X$, $zPx \Rightarrow zPy$ and $\exists w \in X$ such that wPy and not wPx. Let $x = (x_1, \ldots, x_n) \in T^n$

$$K_{n/2} = \{x \mid x_i \in \{0, 1\}, \quad i = 1, \dots, n, \quad |\{i \in N \mid x_i = 1\}| \geq n/2\}.$$

Let

$$\widehat{K}_{n/2} = \{x \in K_{n/2} \mid |\{i \in N \mid x_i = 1\}| = [[n/2]] + 1\}.$$

Let

$$K_{n/2}^* = \{x \mid |\{i \in N \mid x_i = 1\}| \geq n/2\}.$$

As with the previous definition of covering, we permit x to be weakly preferred to y.

Theorem 5.59 *Assume $\widehat{K}_{n/2} \subseteq X$. Then $UC_2(R) = K_{n/2}^* \cap X$.*

Proof Let $x \in K_{n/2}^* \cap X$. Suppose $X \backslash K_{n/2}^* \neq \emptyset$. Let $y \in X \backslash K_{n/2}^*$. Then xRy since at least $n/2$ of the components of x equal 1. Also $\forall z \in X, zPx \Rightarrow zRy$ vacuously since $\nexists z \in X$ such that zPx. Since $y \notin K_{n/2}^*$, the number of components of y less than 1 is more than $n/2$. Let w be such that $w_i = 1$ if $y_i < 1$ for $[[n/2]] + 1$ of these y_i and $w_j = 0$ otherwise. Then $w \in \widehat{K}_{n/2} \subseteq X$. Now not wPx and wPy. Hence xC_2y. Thus $y \notin UC_2(R)$. Hence $UC_2(R) \cap (X \backslash K_{n/2}^*) = \emptyset$. If $X \backslash K_{n/2}^* = \emptyset$, then clearly $UC_2(R) \cap (X \backslash K_{n/2}^*) = \emptyset$. Hence $UC_2(R) \subseteq K_{n/2}^* \cap X$.

Let $x \in K_{n/2}^* \cap X$. Then $\nexists w \in X$ such that wPx. Hence $\forall y \in X, \nexists w \in X$ such that not wPy and wPx. Thus $\forall y \in X$, not yC_2x. Hence $x \in UC_2(R)$. Thus $K_{n/2}^* \cap X \subseteq UC_2(R)$.

Corollary 5.60 *Assume $K_{n/2}^* \subseteq X$. Then $UC_2(R) = K_{n/2}^*$.*

Lemma 5.61 *vC_2y and $x \succ v$ implies xC_2y.*

Proof $vRy \Rightarrow xRy$. $zPx \Rightarrow zPv \Rightarrow zPy$. ($wPy$ and not wPv) $\Rightarrow$ (wPy and not wPx). $\blacksquare$

The following example shows that C_2 is not Pareto efficient.

Example 5.62 Let $X = \langle\{(1, 0.5, 0.25), (0.5, 0.25, 1), (0.25, 1, 0.5)\}\rangle$.
  Consider the triple $(a, 0.5, 0.25)$, where $a \in \{0.75, 1\}$.
  Now $(0.25, 1, 0.5)R(a, 0.5, 0.25)$, but

$$(0.5, 0.25, 1)P(0.25, 1, 0.5) \nRightarrow (0.5, 0.25, 1)P(a, 0.5, 0.25).$$

Thus not $(0.25, 1, 0.5)C_2(a, 0.5, 0.25)$. Also, it's not the case that

$$(0.5, 0.25, 1)R(a, 0.5, 0.25).$$

Thus not $(0.5, 0.25, 1)C_2(a, 0.5, 0.25)$. By Lemma 5.61, it follows that $(a, 0.5, 0.25)$ is C_2-uncovered. Note that $(0.75, 0.5, 0.25)$ is a descendant of $(1, 0.5, 0.25)$ and that $K_{n/2}^* \cap X = \emptyset$.

5.5.3.2 More on C_2

We now show that the Fishburn set, C_2, has Pareto deficient .outcomes. Therefore, the C_2 definition of covering is problematic.

Let $X \subseteq T^3$ and let $I_1 = \{(1, 0, 0), (0, 0, 1), (0, 1, 0)\}$.

Definition 5.63 Define the binary relation C_2 on X by $\forall x, y \in X$, $xC_2y \Leftrightarrow xRy$ and $\forall z \in X[yPz \Rightarrow xPz]$ and $\exists w \in X[xPw$ and not $yPw]$.

Proposition 5.64 Let $x, y \in X$. If xC_2y, then either $x \succ y$ or $y \in \langle I_1 \rangle$.

Proof Suppose $x \preceq y$. Then clearly $\nexists w \in X$ such that xPw and not yPw, a contradiction since xC_2y. Thus either $x \succ y$ or x and y are not comparable with respect to $\preceq$. Suppose x and y are not comparable w.r.t. $\preceq$. Then, without loss of generality, we may assume $x = (x_1, x_2, x_3)$ and $y = (y_1, y_2, y_3)$, where $x_2 > y_2$, and $x_3 < y_3$. Suppose $\exists y_1' \in X$ such that $y_1' < y_1$. Let $z = (y_1', x_2, x_3)$. Then yPz, but not xPz, a contradiction. Hence no such y_1' exists. Thus $y_1 = 0$. Hence $y = (0, y_2, y_3)$. Suppose $y_2 > 0$. Let $z = (x_1, 0, x_3)$. Then yPz, but not xPz, a contradiction. Hence $y_2 = 0$. Thus $y \in \langle I_1 \rangle$.

Proposition 5.65 Let $x, y \in X$. Suppose $x \succ y$. Then either $\exists w \in X[xPw$ and not $yPw]$ or $x \in \langle I_1 \rangle$.

Proof Suppose x and y differ in two or more components. Then let $w = y$. Suppose x and y differ in exactly one component. Then there is no loss in generality in assuming that $x = (x_1, x_2, x_3)$ and $y = (x_1, x_2, y_3)$, where $y_3 < x_3$. Suppose that $x_1 > 0$ or $x_2 > 0$, say $x_2 > 0$. Then $\exists w_2$ such that $0 \leq w_2 < x_2$. Let $w = (x_1, w_2, y_3)$. Then xPw since $x_2 > w_2$ and $x_3 > y_3$, but not yPw since only $x_2 > w_2$. If $x_1 = x_2 = 0$, then $x \in \langle I_1 \rangle$.

Corollary 5.66 Let $x, y \in X$. Suppose $x \succ y$. If $x \in PS_N(R)$, then either $\exists w \in X[xPw$ and not $yPw]$ or $x \in I_1$.

Proposition 5.67 Let $x, y \in X$. If $x \succ y$, then $\forall z \in X$ $[yPz$ implies $xPz]$.

Proposition 5.68 Let $x, y \in X$. If $x \succ y$, then xC_2y or $x \in \langle I_1 \rangle$.

Proof The proof follows from Propositions 5.65 and 5.67 and the fact that $x \succ y$ implies xRy.

Proposition 5.69 Let $x, y \in X$. Suppose $x, y \notin \langle I_1 \rangle$. Then $xC_2y \Leftrightarrow x \succ y$.

Proof The proof follows from Propositions 5.64 and 5.68.

Definition 5.70 Let $x \in X$. Then x is C_2-uncovered $\Leftrightarrow \nexists y \in X$ such that yC_2x.

Proposition 5.71 Let $x \in X$. Suppose x is C_2-uncovered. Then either $x \in PS_N(R)$ or $x \in \langle I_1 \rangle$.

Proof Suppose $x \notin PS_N(R)$. Then $\exists y \in X$ such that $y_i > x_i$ for some i, but $\nexists j$ such that $x_j > y_j$, where $x = (x_1, x_2, x_3)$ and $y = (y_1, y_2, y_3)$. Thus $y \succ x$. Hence yRx. Also $\forall z \in X, xPz \Rightarrow yPz$. We may assume without loss of generality that $x_1 \le y_1, x_2 \le y_2$, and $x_3 < y_3$. If $x_1 = x_2 = 0$, then $x \in \langle I_1 \rangle$. Suppose either $x_1 > 0$ or $x_2 > 0$, say $x_2 > 0$. Then $\exists x_2'$ such that $0 \le x_2' < x_2$. Thus for $w = (x_1, x_2', x_3)$, we have that yPw and not xPw. Hence yC_2x, a contradiction of the hypothesis. Thus $x \in PS_N(R)$.

Example 5.72 Let $PS_N(R) = \{(1, 0.25, 0), (0.25, 0, 1), (0, 1, 0.25)\}$ and $X = \langle PS_N(R) \rangle$. Let $x = (0, 0, 1)$. Then $\forall y \in X, \forall z \in X, xPz$ implies yPz vacuously since $\nexists z \in X$ such that xPz. Let $y = (1, 0.25, 0)$. Then yPx and so yRx. Let $w = (0.75, 0, 0)$. Then yPw and not xPw. Thus yC_2x. Also $x \notin PS_N(R)$ and $x \in I_1$.

Now let $x = (25, 0, 0.75)$. Let $y = (0.25, 0, 1)$. Then yRx and $\forall z \in X[xPz \Rightarrow yPz]$. Let $w = (0, 0, 0.75)$. Then yPw and not xPw. Thus yC_2x. Hence it looks like $PS_N(R)$ is the C_2-uncovered set.

Definition 5.73 Define the binary relation C on X by $\forall x, y \in X, xCy \Leftrightarrow xPy$ and $\forall z \in X[yPz \Rightarrow xPz]$.

Let $UC(R)$ denote the uncovered set of X with respect to R. Let $S_{UC(R)}$ denote the unique minimal generating set of $UC(R)$. Let $UC_2(R)$ denote the set of all C_2-uncovered elements.

Proposition 5.74 $S_{UC(R)} \subseteq UC_2(R) \cup I_1$.

Proof Let $y \in S_{UC(R)}$. If $\exists x \in X$ such that xC_2y, then $x \succ y$ or $y \in \langle I_1 \rangle$ by Proposition 5.64. Since $y \in S_{UC(R)}$, $\nexists x \in X$ such that $x \succ y$. Hence $y \in \langle I_1 \rangle$. Since $\nexists x \in X$ such that $x \succ y, y \in I_1$.

Proposition 5.75 $UC_2(R) \subseteq S_{UC_2(R)} \cup \langle I_1 \rangle$.

Proof Let $y \in UC_2(R)$. Then $\exists x \in S_{UC_2(R)}$ such that $y \in \langle \{x\} \rangle$. Hence $y = x$ or $y \prec x$. If $y = x$, then of course $y \in S_{UC_2(R)}$. Suppose $y \prec x$. Then xRy (or possibly even xPy) and clearly $\forall z \in X, yPz \Rightarrow xPz$. By Proposition 5.65, either $\exists w \in X[xPw$ and not $yPx]$ or $x \in \langle I_1 \rangle$. If such a w exists, then xC_2y, contradicting $y \in UC_2(R)$. Thus $x \in \langle I_1 \rangle$.

Example 5.76 Let $PS_N(R) = I_1$ and $X = \langle I_1 \rangle$. Let $x, y \in X$. Then $\nexists w \in X$ such that xPw and not yPw since X has no strict preferences. Thus $UC_2(R) = \emptyset$ and $UC(R) = X$.

Theorem 5.77 $S_{UC(R)} \cup \langle I_1 \rangle = S_{UC_2(R)} \cup \langle I_1 \rangle$.

Proof By Propositions 5.74 and 5.75, $S_{UC(R)} \subseteq UC_2(R) \cup I_1 \subseteq S_{UC_2(R)} \cup \langle I_1 \rangle$. Thus it follows that $S_{UC(R)} \subseteq S_{UC_2(R)} \cup I_1$. Let $y \in S_{UC_2(R)}$. Suppose $y \notin S_{UC(R)}$. Then $y \notin \langle S_{UC(R)} \rangle$ since $\nexists x \in X$ such that $x \succ y$. Thus y is covered. Thus $\exists x \in X$ such that xPy and $\forall z \in X, yPz$ implies xPz, i.e., xCy. Since not $x \succ y$ and xPy, one of the components of x is $<$ the corresponding component of y and two of the components of x are $>$ the corresponding components of y, say $x = (x_1, x_2, x_3)$ and $y = (y_1, y_2, y_3)$ with $x_1 < y_1$ and $x_2 > y_2$, $x_3 > y_3$. If $y_2 > 0$, then $yP(x_1, 0, x_3)$, but not $xP(x_1, 0, x_3)$. Thus not xCy, a contradiction. Hence $y_2 = 0$ and similarly $y_3 = 0$. Thus $y \in \langle I_1 \rangle$. Hence $S_{UC_2(R)} \subseteq S_{UC(R)} \cup \langle I_1 \rangle$. Thus $S_{UC(R)} \cup \langle I_1 \rangle = S_{UC_2(R)} \cup \langle I_1 \rangle$.

Proposition 5.78 $UC_2(R) \cap \langle I_1 \rangle = \emptyset$.

Proof Let $x \in \langle I_1 \rangle$. Let $y \in X$ be such that two of y's components are positive. Now yRx and $\forall z \in X$, xPz implies yPz vacuously since $\nexists z \in X$ such that xPz. Now $\exists w \in X$ such that yPw and not xPw, namely $w = \bar{0}$. Thus yC_2x. Hence $x \notin UC_2(R)$.

Corollary 5.79 $UC_2(R) = S_{UC_2(R)}$.

Proof The proof follows by Propositions 5.75 and 5.78.

Corollary 5.80 $S_{UC(R)} \cup I_1 = UC_2(R) \cup I_1$.

Proof $UC_2(R) \subseteq S_{UC(R)} \subseteq UC_2(R) \cup I_1$, where the latter inclusion follows since $S_{UC(R)} \cap \langle I_1 \rangle \subseteq I_1$. Thus $UC_2(R) \cup I_1 \subseteq S_{UC(R)} \cup I_1 \subseteq UC_2(R) \cup I_1$ from which the desired result is immediate.

Corollary 5.81 $UC(R) \backslash \langle I_1 \rangle = \langle UC_2(R) \rangle$.

Proof Since

$$\langle S_{UC(R)} \backslash I_1 \rangle = \bigcup_{s \in S_{UC(R)} \backslash I_1} \langle \{s\} \rangle$$

$$= \bigcup_{s \in S_{UC(R)}} \langle \{s\} \rangle \backslash \bigcup_{s \in I_1} \langle \{s\} \rangle$$

$$= \langle S_{UC(R)} \rangle \backslash \langle I_1 \rangle,$$

the desired result follows from Proposition 5.78 and Corollary 5.80.

5.5.4 C_2-Uncovered Set

This section continues the consideration of the uncovered set produced by the Fishburn set, C_2. Suppose U^*C_2 is the C_2-uncovered set for $\langle X \rangle$. Suppose that X has holes, i.e., $X \subset \langle X \rangle$. Note that $U^*C(R) \subseteq X$ since $U^*C(R) \subseteq PS_N(R)$. Suppose $w \in \langle X \rangle \backslash X$ and that for $x \in U^*C_2(R)$, xPw and not yPw for some $y \in X$, but no other w exists. Then x is not in $UC_2(R)$, the C_2-uncovered set for X. However $x \in U^*C(R)$ the uncovered set for X.

Proposition 5.82 *Let $x, y \in X$. If xC_2y, then either $x \succ y$ or $y \in \langle J_{n/2} \rangle$.*

Proof Suppose $x \preceq y$. Then clearly $\nexists w \in X$ such that xPw and not yPw, contradiction since xC_2y. Thus either $x \succ y$ or x and y are not comparable with respect to $\preceq$. Suppose x and y are not comparable w. r. t. $\preceq$. Then $\exists i, j \in N$ such that $x_i > y_i$ and $x_j < y_j$. Since xC_2y, xRy. Thus not yPx. Hence strictly fewer that $[[n/2]] + 1$ of the y_i are strictly greater then the corresponding x_i. There is no loss in generality in assuming $y_1 \leq x_1, \ldots, y_r \leq x_r$ and $y_{r+1} > x_{r+1}, \ldots, y_n > x_n$, where $n - r < [[n/2]] + 1$. Suppose $y_1 = \cdots = y_{s-1} = 0$ for $s \geq 1$. (The case $s = 1$ says no $y_i = 0$.) We show $s - 1 \geq n/2$. Assume $s - 1 < n/2$. Then $n - s + 1 > n/2$. Let $x = (x_1, \ldots, x_n)$ and $y = (y_1, \ldots, y_n)$. Let t be such that $r - t - s + 1 + n - r = [[n/2]] + 1$. Now let $y'_s, \ldots, y'_{r-t}$ be such that $y_s > y'_s \geq 0, \ldots, y_{r-t} > y'_{r-t} \geq 0$. Let

$$z = (x_1, \ldots, x_{s-1}, \overbrace{y'_s, \ldots, y'_{r-t}}^{r-t-s+1}, \overbrace{x_{r-t+1}, \ldots, x_r}^{t}, x_{r+1}, \ldots, x_n).$$

Now $n - t - s + 1 = [[n/2]] + 1$. Thus

$$t + s - 1 = \begin{cases} [[n/2]] - 1 & \text{if } n \text{ is even,} \\ [[n/2]] & \text{if } n \text{ is odd.} \end{cases}$$

Now $t+s-1$ is the number of components x_i is strictly greater than the corresponding components of z. Thus not xPz. However, $r-t-s+1+n-r = [[n/2]]+1$ (see above) is the number of components of y that are strictly greater than the corresponding components of z. Thus yPz, contradicting the hypothesis that xC_2y. Thus $s-1 \geq n/2$. Hence $y \in \langle J_{n/2} \rangle$.

Proposition 5.83 *Let* $x, y \in X$. *If* $x \succ y$, *then either* ($\exists w \in X$ *such that* xPw *and not* yPw) *or* $x \in \langle J_{n/2} \rangle$.

Proof Suppose x and y differ in $[[n/2]] + 1$ or more components. Then let $w = y$. Suppose x and y differ in fewer than $[[n/2]] + 1$ components. There is no loss in generality in assuming that $x = (x_1, \ldots, x_n)$ and $y = (y_1, \ldots, y_n)$, where $x_1 = y_1, \ldots, x_r = y_r, x_{r+1} > y_{r+1}, \ldots, x_n > y_n$, and $n - r \leq [[n/2]]$. Thus

$$r \geq n - [[n/2]] = \begin{cases} [[n/2]] & \text{if } n \text{ is even,} \\ [[n/2]] + 1 & \text{if } n \text{ is odd.} \end{cases}$$

Suppose $x \notin \langle J_{n/2} \rangle$. Then fewer than $n/2$ of the $x_i = 0$, say $x_1 = \cdots = x_s = 0$, where $s < n/2$. Thus $s < r$. Let t be such that $t - s + n - r = [[n/2]] + 1$. Let

$$w = (x_1, \ldots, x_s, \overbrace{w_{s+1}, \ldots, w_t}^{t-s}, \overbrace{y_{t+1}, \ldots, y_r}^{r-t}, \overbrace{y_{r+1}, \ldots, y_n}^{n-r}),$$

where $x_{s+1} > w_{s+1} \geq 0, \ldots, x_t > w_t \geq 0$. Then $t - s + n - r$ of the components of x are strictly greater than the corresponding components of w. Thus xPw. Now $t - s$ components of y are greater than the corresponding components of w and $t - s \leq [[n/2]]$. Hence not yPw.

Proposition 5.84 *Let $x \in X$. Suppose x is C_2-uncovered. Then either $x \in PS_N(R)$ or $x \in \langle J_{n/2} \rangle$.*

Proof Suppose $x \notin PS_N(R)$. Then $y \in X$ such that $y_i > x_i$ for some $i \in N$ and $\nexists j \in N$ such that $x_j > y_j$, where $x = (x_1, \ldots, x_n)$ and $(y_1, \ldots, y_n)$. Thus $y \succ x$. Hence yRx. Also $\forall z \in X, xPz \Rightarrow yPz$. We may assume without loss of generality that $x_1 \leq y_1, \ldots, x_r \leq y_r, x_{r+1} < y_{r+1}, \ldots, x_n < y_n$ and $1 \leq n - r \leq [[n/2]]$. If for $n/2 \leq s \leq r, x_1 = \cdots = x_s = 0$, then $x \in \langle J_{n/2} \rangle$ since

$$r \geq n/2 \left(r \geq n - [[n/2]] = \begin{cases} [[n/2]] & \text{if } n \text{ is even,} \\ [[n/2]] + 1 & \text{if } n \text{ is odd} \end{cases} \right).$$

Suppose $x_1 = \cdots = x_s = 0$, where $s < n/2$. Then $\exists x'_{s+1}, \ldots, x'_r$ such that $0 \leq x'_i < x_i i = s+1, \ldots, r$. Let $w = (0, \ldots, 0, x'_{s+1}, \ldots, x'_r, x_{r+1}, \ldots, x_n)$. Then yPw since $n - s$ of the components of y are strictly greater than the corresponding components of w and $n - s > n/2$ since $s < n/2$. Now not xPw since $r - s$ of the components of x are greater than the corresponding components of w and $r - s \leq n/2$ since $n/2 \leq s \leq r$ whence $n - s \leq n/2$. Thus yC_2x, contradicting the hypothesis that x is C_2-uncovered.

Definition 5.85 Let $\widehat{C} : \mathscr{P}^*(X) \to \mathscr{P}^*(X)$. Then $\widehat{C}$ is called a **choice function** on X if $\forall S \in \mathscr{P}^*(X), \widehat{C}(S) \subseteq S$.

We next introduce a binary relation on X induced by a choice function on X.

Definition 5.86 Let $\widehat{C}$ be a choice function on X. Let $R = \{(x, y) \in X \times X \mid x \in \widehat{C}(S), y \in S \text{ for some } S \in \mathscr{P}^*(X)\}$. Then R is called the **binary operation induced** by $\widehat{C}$.

Since $\widehat{C}(\{x\}) = \{x\}$ for all $x \in X, R$ is reflexive. Since $\emptyset \neq \widehat{C}(\{x, y\}) \subseteq \{x, y\}, R$ is complete.

Example 5.87 Let $X = \{x, y, z\}$. Let $\widehat{C}$ be the choice function on X defined as follows:

$$\widehat{C}(\{x\}) = \{x\}, \ \widehat{C}(\{y\}) = \{y\}, \ \widehat{C}(\{x\}) = \{z\},$$
$$\widehat{C}(\{x, y\}) = \{x, y\}, \ \widehat{C}(\{x, z\}) = \{z\}, \ \widehat{C}(\{y, z\}) = \{z\},$$
$$\widehat{C}(\{x, y, z\}) = \{x, y\}.$$

Let R be the binary relation on X induced by $\widehat{C}$. Then

$$R = \{(x, x), (y, y), (z, z), (x, y), (y, x), (z, x), (z, y), (x, z), (y, z)\}$$
$$= X \times X.$$

Now $M(R, \{x, z\}) = \{x, z\} \neq \{z\} = \widehat{C}(\{x, z\})$. Thus R does not rationalize $\widehat{C}$.

Example 5.88 Let $X = \{x, y, z\}$. Let $\widehat{C}$ be the choice function on X defined as follows:

$$\widehat{C}(\{x\}) = \{x\}, \; \widehat{C}(\{y\}) = \{y\}, \; \widehat{C}(\{x\}) = \{z\},$$
$$\widehat{C}(\{x, y\}) = \{x, y\}, \; \widehat{C}(\{x, z\}) = \{z\}, \; \widehat{C}(\{y, z\}) = \{z\},$$
$$\widehat{C}(\{x, y, z\}) = \{z\}.$$

Let R be the binary relation on X induced by $\widehat{C}$. Then

$$R = \{(x, x), (y, y), (z, z), (x, y), (y, x), (z, x), (z, y)\}.$$

Now $(x, z) \notin R$ and $(y, z) \notin R$. Thus zPX and zPy. We have

$$M(R, X) = \{z\} = \widehat{C}(X),$$
$$M(R, \{x\}) = \{x\} = \widehat{C}(\{x\}),$$
$$M(R, \{y\}) = \{y\} = \widehat{C}(\{y\}),$$
$$M(R, \{z\}) = \{z\} = \widehat{C}(\{z\}),$$
$$M(R, \{x, y\}) = \{x, y\} = \widehat{C}(\{x, y\}),$$
$$M(R, \{x, z\}) = \{z\} = \widehat{C}(\{x, z\}),$$
$$M(R, \{y, z\}) = \{z\} = \widehat{C}(\{y, z\}).$$

Thus R rationalizes $\widehat{C}$.

Let C_{mgs} be the choice function on X defined by for all nonempty subsets Y of X, $C_{mgs}(Y) = S_Y$, where S_Y is the unique minimal generation set of Y. In the next example, we show that it is possible for $UC(R) \supset S_{UC(R)}$.

Example 5.89 Let $U = T^3$. Let $X \subseteq U$ and R be a binary relation on X such that $PS_N(R) = \{(1, 0.75, 0.5), (0.75, 0.5, 1), (0.5, 1, 0.75)\}$ and $X = \langle PS_N(R) \rangle$. Let $UC(R)$ denote the uncovered set of X with respect to R. Then we show that $C_{mgs}(UC(R)) = \{(1, 0.75, 0.5), (0.75, 0.5, 1), (0.5, 1, 0.75)\}$. It suffices to show that $(1, 0.75, 0.5), (0.75, 0.5, 1), (0.5, 1, 0.75)$ are uncovered since $\nexists \overline{x} \in X$ such that $\overline{x} \succ$ any of these elements. Suppose $(a, b, c)P(1, 0.75, 0.5)$. Then $b = 1$ and $c > 0.5$. Let $\overline{z} = (0.75, 0.5, 1)$. Then $(1, 0.75, 0.5)P\overline{z}$. However not $(a, b, c)P\overline{z}$ else $a = 1$, but then $(a, b, c) \notin X$. Thus $(1, 0.75, 0.5)$ is uncovered. Similarly, $(0.75, 0.5, 1)$ and $(0.5, 1, 0.75)$ are uncovered. We next show that $(0.75, 0.75, 0.5) \in UC(R)$ and so $UC(R) \supset S_{UC(R)}$. Suppose $\overline{x}C(0.75, 0.75, 0.5)$. Then $\overline{x}P(0.75, 0.75, 0.5)$. Thus $x_1 = 1, x_2 = 1$ or $x_1 = 1, x_3 \geq 0.75$ or $x_2 = 1, x_3 \geq 0.75$, where $\overline{x} = (x_1, x_2, x_3)$. The first two cases are impossible. However $(a, 1, 0.75)P(0.75, 0.75, 0.5)$, where $a \leq 0.5$. $((a, 1, 0.5) \notin X$ if $a > 0.5$.) Now $(0.75, 0.75, 0.5)P\overline{z}$, but not $(a, 1, 0.75)P\overline{z}$, where $\overline{z} = (0.5, 0.5, 1)$. Thus $(a, 1, 0.75)$ does not cover $(0.75, 0.75, 0.5)$. Hence $(0.75, 0.75, 0.5) \in UC(R)$.

Let R_{mgs} denote the relation induced by C_{mgs}. Let $Y = \{(0.25, 0, 0)\} \cup \{(0, b, c) \mid (0, b, c) \in X\}$. Then $C_{mgs}(Y) = \{(0.25, 0, 0), (0, 1, 0.75), (0, 0.5, 1)\}$. Thus $\{(0.25, 0, 0), (0, 1, 0.75), (0, 0.5, 1)\} \times Y \subseteq R_{mgs}$.

Example 5.90 Let

$$PS_N(R) = \{(0.25, 0.25, 0.25), (1, 0, 0), (0, 1, 0), (0, 0, 1)\}.$$

Let $X = \langle PS_N(R) \rangle$. Let $x = (0.25, 0.25, 0.25)$ and $y = (0.25, 0.25, 0)$. Then it is easily verified that xCy. Let $W_{x,y} = \{w \in X \mid xPw$ and not $yPw, w \neq (0, 0, 0)\}$. Then $W_{x,y} = \langle \{(1, 0, 0), (0, 1, 0)\} \rangle \backslash \{(0, 0, 0)\}$. However, $(1, 0, 0)$, $(0, 1, 0)$ cannot be holes. Let $H = \langle \{(0.75, 0, 0,), (0, 0.75, 0)\} \rangle \backslash \{(0, 0, 0))\}$. Then xCy in $X \backslash H$ since $(1, 0, 0), (0, 1, 0) \in X \backslash H$. Note that not xCy in $X \backslash W_{x,y}$.

Example 5.91 Let

$$PS_N(R) = \{(0.25, 0.25, 0.25), (1, 1, 0), (0, 01)\}.$$

Let $X = \langle PS_N(R) \rangle$. Let $x = (0.25, 0.25, 0.25)$ and $y = (0.25, 0.25, 0)$. Then it is easily verified that xCy. Let $W_{x,y} = \{w \in X \mid xPw$ and not $yPw, w \neq (0, 0, 0)\}$. Then $W_{x,y} = \langle \{(1, 0, 0), (0, 1, 0)\} \rangle \backslash \{(0, 0, 0)\}$. In this example, $(1, 0, 0)$ and $(0, 1, 0)$ can be holes. Thus not xCy in $X \backslash W_{x,y}$.

5.5.5 Covering Relation C_7

We next combine C_2 and C_4, which results in a definition of covering used by McKelvey (1986) that we label C_7. When indifference is not permitted, C_2, C_4, and C_7 result in identical uncovered sets. C_7 In what follows, we demonstrate that the same is not true when we introduce thick indifference.

Definition 5.92 Let $x, y \in X$. Define the binary operation C_7 on X as follows: $\forall x, y \in X, xC_7y \Leftrightarrow ([\forall z \in X, zPx \Rightarrow zPy$ and $\exists w \in X, wPy$, not $wPx]$ and $[\forall z \in X, yPz \Rightarrow xPz])$ or $([\forall z \in X, zPx \Rightarrow zPy]$ and $[\forall z \in X, yPz \Rightarrow xPz$ and $\exists w \in X, xPw$, not $yPw])$.

Proposition 5.93 *Assume* $\forall v \in X, \nexists z \in X$ *such that* vPz *and* $\nexists z \in X$ *such that* zPv. *Then* $UC_7(R) = X$.

Proof Let $x, y \in X$. By hypothesis, $\forall z \in X, zPx$ implies zPy and $\forall z \in X, yPz$ implies xPz hold vacuously. Thus $xC_7y \Leftrightarrow (\exists w \in X, wPy$, not $wPx)$ or $(\exists w \in X, xPw$, not $yPw)$. By hypothesis, $\nexists w \in X$, such that wPy and $\nexists w \in X$ such that xPw. Thus not xC_7y. Since x and y are arbitrary, no element of X is covered. Hence $UC_7(R) = X$.

Let $U = T^n$. Let $I_{n/2} = \{x = (x_1, \ldots, x_n) \mid x_i \in \{0, 1\}$ and $1 \leq |\{i \in N \mid x_i = 1\}| \leq n/2\}$.

Example 5.94 Let $U = T^n$. Let $X \subseteq \langle I_{n/2} \rangle$. Then $\forall v \in X, \nexists z \in X, vPz$ and $\nexists z \in X, zPv$. Thus by Proposition 5.93, $UC_7(R) = X$. (In our situation, $X \supseteq \{x = (x_1, \ldots, x_n) \mid x_i \in \{0, 1\}$ and $|\{i \in N \mid x_i = 1\}| = 1\}$.) Note that there are descendants with respect to the partial order $\leq$ on T^n.

Example 5.95 Let $U = T^3$. Let $X = \langle I_{\frac{3}{2}} \rangle \cup \{(0.25, 0.25, 0)\}$. We show that $UC_7(R) = \{(0.25, 0.25, 0)\}$.

Let $x \in \langle \{(1, 0, 0), (0, 1, 0)\} \rangle$. Then $\exists w \in X$ such that $(0.25, 0.25, 0)Pw$ and not xPw, namely $w = (0, 0, 0)$. Also $\forall z \in X, zP(0.25, 0.25, 0) \Rightarrow zPx$ vacuously and $\forall z \in X, xPz \Rightarrow (0.25, 0.25, 0)Pz$ vacuously. Thus $(0.25, 0.25, 0)C_7\bar{i}$.

Let $x \in \langle \{(0, 0, 1)\} \rangle$. Then $\exists w \in X$ such that $wP(0, 0, 1)$ and not $wP(0.25, 0.25, 0)$, namely $w = (0.25, 0.25, 0)$. Also $\forall z \in X, zP(0.25, 0.25, 0)$ implies zPx vacuously and $\forall z \in X, xPz$ implies $(0.25, 0.25, 0)Pz$ vacuously. Thus $(0.25, 0.25, 0)C_7x$. Then $\exists w \in X$ such that $wP(0, 0, 1)$ and not $wP(0.25, 0.25, 0)$, namely $w = (0.25, 0.25, 0)$.

Also $\forall z \in x, z P(0.25, 0.25, 0)$ implies zPx vacuously and $\forall z \in X, xPz$ implies $(0.25, 0.25, 0)Px$ vacuously. Now let $x \in \langle I_1 \rangle$. Then $\nexists w \in X$ such that xPw and not $(0.25, 0.25, 0)Pw$ and $\nexists w \in X$ such that $wP(0.25, 0.25, 0)$ and not wPx. Hence not $xC_7(0.25, 0.25, 0)$. Thus $UC_7(R) = \{(0.25, 0.25, 0)\}$. By Example 5.94, $UC_7(R) = \langle I_{\frac{3}{2}} \rangle$ if $X = \langle I_{\frac{3}{2}} \rangle$.

Example 5.96 Let $U = T^4$ and $X = \langle I_2 \rangle \cup \{(0.25, 0.25, 0.25, 0)\}$. We show that $UC_7(R) = \{(0.25, 0.25, 0.25, 0)\}$.

Let $x \in \langle I_2 \backslash \{(0, 0, 0, 1)\} \rangle$. Then there $\exists w \in X$ such that $(0.25, 0.25, 0.25, 0)Pw$ and not xPw, namely $w = (0, 0, 0, 0)$. Also $\forall z \in X, zP(0.25, 0.25, 0.25, 0) \Rightarrow zPx$ vacuously and $\forall z \in X, xPz$ implies $(0.25, 0.25, 0.25, 0)Pz$ vacuously. Thus $(0.25, 0.25, 0.25, 0)C_7x$.

Let $x \in \langle \{(0, 0, 0, 1)\} \rangle$. Then there $\exists w \in X$ such that wPx and not $wP(0.25, 0.25, 0.25, 0)$, namely $w = (0.25, 0.25, 0.25, 0)$. Also $\forall z \in X, zP(0.25, 0.25, 0.25, 0) \Rightarrow zPx$ vacuously and $\forall z \in X, xPz$ implies $(0.25, 0.25, 0.25, 0)Pz$ vacuously. Thus $(0.25, 0.25, 0.25, 0)C_7x$.

Now let $x \in \langle I_2 \rangle$. Then there $\nexists w \in X$ such that xPw and not $(0.25, 0.25, 0.25, 0)Pw$ and $\nexists w \in X$ such that $wP(0.25, 0.25, 0.25, 0)$ and not wPx. Hence it is not the case that $xC_7(0.25, 0.25, 0.25, 0)$. Thus $UC_7(R) = \{(0.25, 0.25, 0.25, 0)\}$. By Example 5.94, $UC_7(R) = \langle I_2 \rangle$ if $X = \langle I_2 \rangle$.

Let

$$H_{n/2} = \{x = (x_1, \ldots, x_n) \in X \mid |\{i \in N \mid x_i = 1\}| \geq [[n/2]] + 1\} \ .$$

Theorem 5.97 *Let $x, y \in X$. Then $x \succ y$ implies either xC_7y or*

$$\left(x \in \langle J_{n/2} \rangle \text{ and } \nexists w \in X \text{ such that } wPy \right) \ .$$

Proof Suppose $x \succ y$. Then $\forall z \in X, yPz$ implies xPz and $\forall z \in X, zPx$ implies zPy. Suppose not xC_7y. Then not $(\exists w \in X$ such that xPw and not yPw or $\exists w \in X$ such that wPy and not $wPx)$. Thus not $(\exists w \in X$ such that xPw and not $yPw)$ and not

($\exists \in X$ such that wPy and not wPx). Hence $\forall w \in X$, not (xPw and not yPw) and $\forall w \in X$, not (wPy and not wPx). Thus ($\forall w \in X$, not xPw or yPw) and ($\forall w \in X$, not wPy or wPx). Hence $\forall w \in X, xPw \Rightarrow yPw$ and $\forall w \in X, wPy \Rightarrow wPx$. It suffices to assume x and y differ in one component. Suppose $\exists w \in X$ such that xPw. Then x has $[[n/2]] + 1$ or more components than the corresponding components of w, say $w = (w_1, \ldots, w_r, x_{r+1}, \ldots, x_n)$, where $x_1 > w_1, \ldots, w_r > x_2$ and $r \geq [[n/2]] + 1$. Let $w' = (w_1, \ldots, w_s, x_{s+1}, \ldots, x_n)$, where $s = [[n/2]] + 1$. If $x_i > y_i$, where $s + 1 \leq i \leq n$, then let $w'' = (w_1, \ldots, w_{s-1}, x_s, ..x_{i-1}, y_i, x_i, \ldots, x_n)$. If $x_i > y_i$, where $1 \leq i \leq s$, then let $w'' = (w_1, \ldots, w_{i-1}, y_i, w_{i+1}, \ldots, w_s, x_{s+1}, \ldots, x_n)$. Then in either case xPw, but not yPw, a contradiction. Thus $\nexists w \in X$ such that xPw. Hence, not $xP(0, \ldots, 0)$. Thus x has fewer than $[[n/2]] + 1$ nonzero components. Hence $x \in \langle J_{n/2} \rangle$. Now suppose $\exists w \in X$ such that wPy, say $w = (w_1, \ldots, w_r, y_{r+1}, \ldots, y_n)$, where $w_1 > y_1, \ldots, w_r > y_r$ and $r \geq [[n/2]] + 1$. Let $w' = (w_1, \ldots, w_s, y_{s+1}, \ldots, y_n)$, where $s = [[n/2]] + 1$. If $x_i > y_i$, where $s+1 \leq i \leq n$, then let $w'' = (w_1, \ldots, w_{s-1}, y_s, \ldots, y_{i-1}, x_i, y_{i+1}, \ldots, y_n)$. If $x_i > y_i$, where $1 \leq i \leq s$, then let $w'' = (w_1, \ldots, w_{i-1}, x_i, w_{i+1}, \ldots, w_s, y_{s+1}, \ldots, y_n)$. In either case, $w''Py$, but not $w''Px$, a contradiction. Thus $\nexists w \in X$ such that wPy.

Note that if $y \in H_{n/2}$, then $\nexists w \in X$ such that wPy.

Proposition 5.98 Let $x, y \in X$. Suppose xPy and $y \in \langle J_{n/2} \rangle$. Then xC_7y if and only if $\forall z \in X, zPx$ implies zPy.

Proof $\forall z \in X, yPz$ implies xPz vacuously since $\nexists z \in X$ such that yPz due to the fact that $y \in \langle J_{n/2} \rangle$. Also, $\exists w \in X$ such that xPw and not yPw, namely $w = y$. The conclusion now follows from the definition of C_7.

Example 5.99 Let $U = T^3$. Let $X = \langle \{(, 25, 0.25, 0), (0, 0.5, 0.25)\} \rangle \cup I_{\frac{3}{2}}$. We first show that $(0, 0.5, 0.25)C_7(0, 0, 1)$. Let $z \in X$. Then $zP(0, 0.5, 0.25)$ implies $zP(0, 0, 1)$ vacuously and $(0, 0, 1)Pz$ implies $(0, 0.5, 25)Pz$ vacuously. $\exists w \in X$ such that $(0, 0.5, 0.25)Pw$ and not $(0, 0, 1)Pw$, namely $w = (0, 0.25, 0)$. Also, $\exists w \in X$ such that $wP(0, 0, 1)$ and not $wP(0, 0.5, 0.25)$, namely $w = (0.25, 0.25, 0)$. Thus $(0, 0.5, 0.25)C_7(0, 0, 1)$. Note that

$$\text{not } (0, 0.5, 0.25) \succ (0, 0, 1) \text{ and not } (0, 0.5, 0.25)P(0, 0, 1).$$

We next note that
$$\text{not } (0.25, 0.25, 0)C_7(0, 0, 1).$$

This follows by Proposition 5.98 since $(0, 0.5, 0.25)P(0.25, 0.25, 0)$, but it is not the case that $(0, 0.5, 0)P(0, 0, 1)$. We next show that

$$(0, 0.5, 0.25)C_7(0, 1, 0).$$

We have that $zP(0, 0.5, 0.25)$ implies that $zP(0, 1, 0)$ vacuously and $(0, 1, 0)Pz$ implies $(0, 0.5, 25)Pz$ vacuously. $\exists w \in X$ such that $(0, 0.5, 0.25)Pw$ and not $(0, 1, 0)Pw$, namely $w = (0, 0.25, 0)$. Thus $(0, 0.5, 0.25)C_7(0, 1, 0)$.

Also $zP(0, 0.5, 0.25) \Rightarrow zP(1, 0, 0)$ vacuously and $(1, 0, 0)Pz \Rightarrow (0, 0.5, 25)Pz$ vacuously. $\exists w \in X$ such that $(0, 0.5, 0.25)Pw$ and not $(1, 0, 0)Pw$, namely $w = (0, 0.25, 0)$. Thus

$$(0, 0.5, 0.25)C_7(1, 0, 0).$$

We now show that not

$$(0, 0.5, 0.25)C_7(0.25, 0.25, 0)$$

even though $(0, 0.5, 0.25)P(0.25, 0.25, 0)$. We have that $zP(0, 0.5, 0.25) \Rightarrow zP(0.25, 0.25, 0)$ vacuously, but $(0.25, 0.25, 0)Pz \nRightarrow (0, 0.5, 0.25)Pz$ for $z = (0, 0, z_3)$ with $z_3 \geq 0.25$. Further routine calculations show that

$$UC_7(R) = \{(0, 0.5, 0.25), (0.25, 0.25, 0)\}.$$

Since $(0, 0.5, 0.25)P(0.25, 0.25, 0)$, it follows that

$$M(R, X) = \{(0, 0.5, 0.25)\}.$$

Definition 5.100 Let Y be a subspace of X. Then Y is called **primary** (Mordeson and Nair 1996) if the following conditions hold:

(i) $\exists y \in X$ such that $Y = \langle y \rangle$;
(ii) $\forall x \in X$, if $Y \subseteq \langle x \rangle$, then $Y = \langle x \rangle$.

Recall that S_X denotes the unique minimal generating set of X. It follows that $\{\langle x \rangle \mid x \in S_X\}$ is the set of all primary subspaces of X. Recall also that $S_X = PS_N(R)$.

Proposition 5.101 *If for all $y \in J_{n/2}$, $\langle y \rangle$ is not primary, then $UC_7(R) \subseteq S_X$.*

Proof Let $y \in J_{n/2}$. By Mordeson and Nair (1996), Corollary 15, p. 121, $\langle y \rangle$ is contained in a primary subspace of X, say $\langle x \rangle$. Since $\langle y \rangle$ is not primary, $\langle y \rangle \subset \langle x \rangle$. Thus $x \succ y$ and $x \notin J_{n/2}$. Thus by Proposition 5.98, xC_7y. Since $\forall y \in X \backslash (S_X \cup J_{n/2})$, $\exists x \in X$ such that $x \succ y$, it follows by Proposition 5.98 that xC_7y. Hence the desired result holds.

5.5.6 A Symmetrical Covering Relation

We have examined four versions of covering definitions modified to permit x to be weakly preferred to y. Only C_4 gives the desired results. We now consider a definition of covering that permits x to cover y and y to cover x. We demonstrate that this has an undesirable result: there are Pareto deficient elements in the uncovered set.

Definition 5.102 Define the binary relation C_s on X by $\forall x, y \in X$, xCy if and only if $x \neq y$ and $\forall z \in X$, $yPz \Rightarrow xPz$.

Remark 5.103 $\forall z \in X, yPz \Rightarrow xPz$ implies xRy.

Proof Suppose not xRy. Then since R is complete, yPx. Thus xPx, a contradiction. Thus xRy.

Proposition 5.104 *Let* $x, y \in X$. *If* $x \succ y$, *then* xCy.

Proof Suppose yPz, where $z \in X$. Since $x \succ y$, xPz.

Corollary 5.105 $UC(R) \subseteq S_X$.

Proof $\forall y \in X \backslash S_X$, there exists $x \in S_X$ such that $x \succ y$.

Theorem 5.106 *Let* $x, y \in X$. *If* xCy, *then either* $x \succ y$ *or* $y \in \langle J_{n/2} \rangle$.

Proof Clearly, $y \neq x$. Suppose $x \prec y$. Since xCy, so $\forall z \in X$, yPz implies xPz Since $x \prec y$, we may write with out loss of generality $x = (x_1, \ldots, x_n)$ and $y = (y_1, \ldots, y_n)$ with $x_1 = y_1, \ldots, x_r = y_r, \ldots, x_{r+1} < y_{r+1}, \ldots, x_n < y_n$. Now $n - r < [[n/2]] + 1$ else yPx, but not xPx, a contradiction. If $y_1 = \cdots = y_s = 0$ for $s > [[n/2]]$, then $y \in \langle J_{n/2} \rangle$. Suppose $y \notin \langle J_{n/2} \rangle$. Let

$$z = (0, \overset{s-1}{\ldots}, 0, x'_s, \ldots, x'_{r-t}, x_{r-t+1}, \ldots, x_r, \ldots, x_n),$$

where $0 \leq x'_i < x_i, i = s, \ldots, r - t$, where if $s = 1$, we mean that there are no zeros, and where t is such that $n - s + 1 > [[n/2]]$ and so yPz and $r - t - s + 1 < [[n/2]]$ and so not xPz. However this is impossible. Thus either $x \succ y$ or x and y are not comparable with respect to $\preceq$.

Suppose x and y are not comparable with respect to $\preceq$. Then $\exists i, j \in N$ such that $x_i > y_i$ and $x_j < y_j$. Since xC_2y, xRy. Thus not yPx. Hence strictly fewer that $[[n/2]] + 1$ of the y_i are strictly greater then the corresponding x_i. There is no loss in generality in assuming $y_1 \leq x_1, \ldots, y_r \leq x_r$ and $y_{r+1} > x_{r+1}, \ldots, y_n > x_n$, where $n - r < [[n/2]] + 1$. Suppose $y_1 = \cdots = y_{s-1} = 0$ for $s \geq 1$. (The case $s = 1$ says no $y_i = 0$.) We show $s - 1 \geq n/2$.

Assume $s - 1 < n/2$. Then $n - s + 1 > n/2$. Let $x = (x_1, \ldots, x_n)$ and $y = (y_1, \ldots, y_n)$. Let t be such that $r - t - s + 1 + n - r = [[n/2]] + 1$. Now let $y'_s, \ldots, y'_{r-t}$ be such that $y_s > y'_s \geq 0, \ldots, y_{r-t} > y'_{r-t} \geq 0$. Let

$$z = (x_1, \ldots, x_{s-1}, y'_s, \overset{r-t-s+1}{\ldots}, y'_{r-t}, x_{r-t+1}, \overset{t}{\ldots}, x_r, x_{r+1}, \ldots, x_n).$$

Now $n - t - s + 1 = [[n/2]] + 1$. Thus

$$t + s - 1 = \begin{cases} [[n/2]] - 1 & \text{if } n \text{ is even,} \\ [[n/2]] & \text{if } n \text{ is odd.} \end{cases}$$

Now $t + s - 1$ is the number of components x_i is strictly greater than the corresponding components of z. Thus not xPz. However,

$$r - t - s + 1 + n - r = [[n/2]] + 1$$

(see above) is the number of components of y that are strictly greater than the corresponding components of z. Thus yPz, contradicting the hypothesis that xCy. Thus $s - 1 \geq n/2$. Hence $y \in \langle J_{n/2} \rangle$.

Example 5.107 Let $n = 3$ and

$$X = \langle \{(1, 0, 0), (0, 0, 1), (0, 1, 0)\} \rangle.$$

Then $\forall z \in X, (1, 0, 0)Pz \Rightarrow (0, 0, 1)Pz$ vacuously since $\nexists z \in X$ such that $(1, 0, 0)Pz$. Hence $(1, 0, 0)C(0, 0, 1)$ and similarly $(0, 0, 1)C(1, 0, 0)$. In fact, $(0, 0, 0)C(1, 0, 0)$. Thus it follows that $UC(R) = \emptyset$.

5.5.7 The Implication of Missing n-Tuples

Covering relation C_4 gives the best results. It produces an uncovered set $UC(X)$ that with a trivial exception, is contained in the Pareto set. However, its ability to do so rests on the assumption that $\langle PS_N(R) \rangle$ identifies the set of alternatives, X. When immediate predecessors of elements of the Pareto set $PS_N(\overline{\rho})$ are not contained in the set of alternatives, non-Pareto efficient alternatives may be included in the uncovered set. This problem has been labeled "vulnerability to holes" Mordeson et al. (2011).

We now give consideration to this possibility of missing n-tuples. We begin with an example showing how n-tuples can be missing.

Let $\mathbb{R}_+$ denote the set of nonnegative real numbers and $\mathbb{R}_+^2 = \mathbb{R}_+ \times \mathbb{R}_+$. Let $n = 3$ in this section and let the players' preferences be denoted by μ, ν, and σ. Let $x \in \mathbb{R}_+^2$. Suppose $x \in \mu^r \cap \nu^r \cap \sigma^r$ and $x \notin \mu^{r'} \cap \nu^{r'} \cap \sigma^{r'}$ for $r' \geq r, s' \geq s, t' \geq t$, where one of the inequalities is strict. Let $f^* : \mathbb{R}_+^2 \to T^3$ be defined by $f^*(x) = (r, s, t)$. Let $X = f^*(\mathbb{R}_+^2)$. Suppose for example, $\mu^{0.5} \cap \nu^{0.25} \cap \sigma^0 = \mu^{0.25} \cap \nu^{.25} \cap \sigma^0$. Then $(0.25, 0.25, 0) \notin f^*(\mathbb{R}_+^2)$. That is, $(0.25, 0.25, 0)$ is missing. In this case, $X \subset \langle X \rangle$.

Consider the representation of the preferences of three players if the maximal set is empty. Suppose there are six alternatives in the Pareto set:

$$PS_N(R) = \{(1, 0, 0), (0, 1, 0), (0, 0, 1), (0.5, 0.25, 0), (, 25, 0, 0.5), (0, 0.5, 0.25)\}.$$

The first three alternatives are elements of $\langle J_{n/2} \rangle$ and are not strictly preferred to any other alternative. The remaining three alternatives comprise the unique uncovered set:

$$UC(X) = \{(0.5, 0.25, 0), (, 25, 0, 0.5), (0, 0.5, 0.25)\}.$$

Definition 5.108 Let $x, y \in X$. Suppose xCy. Let $W_{x,y} = \{w \in X \mid xPw$ and not $yPw, y \neq w \neq (0, 0, 0)\}$. Then (x, y) is said to be **vulnerable to holes** if not xCy in $X \backslash W_{x,y}$.

Proposition 5.109 *Let* $x, y \in X$. *Suppose* xCy. *Then* $x \notin \langle I_1 \rangle$.

Proof Suppose $x \in \langle I_1 \rangle$. Then there does not exist $w \in X$ such that xPw. Hence not xCy.

Proposition 5.110 *Let* $x, y \in X$. *Suppose* xCy. *Then* (x, y) *is vulnerable to holes if and only if* $y \notin \langle I_1 \rangle$ *and not* xPy.

Proof Suppose (x, y) is vulnerable to holes. Suppose $y \in \langle I_1 \rangle$. Then xCy in $X \backslash W_{x,y}$ since $(0, 0, 0) \in W_{x,y}$ and not $yP(0, 0, 0)$, where the latter condition holds since $x \notin \langle I_1 \rangle$ by the previous Proposition. It is not the case that xPy else xPy in $W_{x,y}$ since $y \in W_{x,y}$ and not yPy, i.e., y is a w.

Conversely, suppose $y \notin \langle I_1 \rangle$ and not xPy. Suppose xCy in $W_{x,y}$. Then there exists $w \in X \backslash W_{x,y}$ such that xPw and not yPw. Hence either $w = (0, 0, 0)$ or $w = y$. Suppose $w = (0, 0, 0)$. Since not $yPw, y \in \langle I_1 \rangle$, a contradiction. Suppose $w = y$. Then xPy, a contradiction. Hence not xCy in $X \backslash W_{x,y}$.

Theorem 5.111 *Let* $x, y \in X$. *Suppose* xCy. *Suppose also that* (x, y) *is vulnerable to holes. Then* $\forall w \in W_{x,y}$, *the following conditions hold:*

(i) *If* $x \not\succ w$, *then there exists a permutation* π *of* $\{1, 2, 3\}$ *such that* $x_{\pi(1)} = y_{\pi(1)} > w_{\pi(1)}, w_{\pi(2)} > x_{\pi(2)} = y_{\pi(2)}$, *and* $x_{\pi(3)} > w_{\pi(3)} \geq y_{\pi(3)}$.

(ii) *If* $x \succ w$, *then* (a) *there exists a permutation* π *of* $\{1, 2, 3\}$ *such that* $x_{\pi(1)} = y_{\pi(1)} > w_{\pi(1)}, x_{\pi(2)} = y_{\pi(2)} = w_{\pi(2)}$, *and* $x_{\pi(3)} > y_{\pi(3)} = w_{\pi(3)}$ *or* (b) *there exists a component of* w *strictly greater than the corresponding component of* y *and there exists a permutation* π *of* $\{1, 2, 3\}$ *such that* $x_{\pi(1)} = y_{\pi(1)} = w_{\pi(1)}, x_{\pi(2)} = y_{\pi(2)} > w_{\pi(2)}$, *and* $x_{\pi(3)} > w_{\pi(3)} > y_{\pi(3)}$.

Conversely, if $w \in X$ *and satisfies either* (1) *or* (2), *then* $w \in W_{x,y}$.

Proof Let $w \in W_{x,y}$. Then $x \neq w$ since xPw. Since $y \notin \langle I_1 \rangle, x \succ y$ by Proposition 5.82. Since xRy and not xPy, two of the components of x equal the two corresponding components of y and the remaining component of x is greater than the corresponding component of y.

Suppose $x \not\succ w$. Since xPw, two of the components of x are strictly greater than the corresponding components of w and the remaining component of w is strictly greater then the corresponding component of x. For simplicity and without loss of generality, we can write either (i) $x_1 = y_1 > w_1, x_2 = y_2 > w_2$ and $w_3 > x_3 > y_3$ or (ii) $x_1 = y_1 > w_1, w_2 > x_2 = y_2$, and $x_3 > y_3, x_3 > w_3$. However (i) does not hold else yPw. Suppose (ii) holds. If $y_3 > w_3$, then yPw, a contradiction. Thus $w_3 \geq y_3$. Hence (1) holds.

Suppose $x \succ w$. Then as in the previous paragraph, we can write without loss of generality, $x_1 = y_1 > w_1, x_2 = y_2 \geq w_2$, and either (i) $y_3 \geq w_3$ or (ii)$w_3 > y_3$.

Suppose (*i*) holds. Then $x_1 = y_1 > w_1$ and since not $yPw, x_2 = y_2 = w_2$ and $x_3 > y_3 = w_3$. In this case, (2) holds. Suppose (*ii*) holds. Then $x_1 = y_1 = w_1, x_2 = y_2 > w_2$, and $x_3 > w_3 > y_3$ since not yPw. In this case, (2) holds.

For the converse, the only possible way $w = (0, 0, 0)$ is if (2*a*) holds, but then $y \in \langle I_1 \rangle$, a contradiction. Clearly $w \neq y$. It is easily verified that xPw and not yPw.

While the problem induced by missing n-tuples is not trivial, its likelihood of occurrence decreases as the number of players increases. Thus, in the absence of a fuzzy maximal set, scholars may wish to consider using the fuzzy uncovered set in fuzzy public choice models. Covering relation C_4 in particular commends itself for use. However, the fuzzy uncovered set should only be used when appropriate institutional settings are being modeled. Since government formation is not among those institutional settings, we will not make use of the fuzzy uncovered set in this book.

References

Austen-Smith, D., Banks, J.: Positive Political Theory I. The University of Michigan Press, Ann Arbor (1999)

Bianco, W., Jeliazkov, I., Sened, I.: The uncovered set and the limits of legislative action. Political Anal **12**, 256–276 (2004)

Bordes, G.: On the possibility of reasonable consistent majoritarian choice: some positive results. J. Theor. Politics **31**, 122–132 (1983)

Brauninger, T.: Stability in spatial voting games with restricted preference maximizing. J. Theor. Politics **19**(2), 173–191 (2007)

Cox, G.W.: The uncovered set and the core. Am. J. Political Sci. **31**, 408–489 (1987)

Diermeier, D., Eraslan, H., Merlo, A.: A structual model of goverment formation. Econometrica **71**(1), 27–70 (2003)

Ehlers, L., Barbera, S.: Free triples, large indifference classes and the majority rule. Soc. Choice. Welfare. **37**(4), 559–574 (2011). http://dx.doi.org/10.1007/s00355-011-0584-8

Enelow, J., Hinich, M.: The Spatial Theory of Voting. Cambridge University Press, Cambridge (1984)

Epstein, E.: Uncovering some subtleties of the uncovered set. Soc. Choice Theory Distrib. Politics **15**, 81–93 (1988)

Fishburn, P.: Condorcet social choice functions. SIAM J. Appl. Math. **33**, 469–489 (1977)

Gehrlein, W.V., Valognes, F.: Condorcet efficiency: a preference for indifference. Soc. Choice Welf. **18**(1), 193–205 (2001)

Gibilisco, M.B., Gowen, A.M., Albert, K.E., Mordeson, J.N., Wierman, M.J., Clark, T.D.: Fuzzy Social Choice. Springer, New York (2014)

Koehler, D.H.: Convergence and restricted preference maxprefere under simple majority rule: results from a computer simulation of committee choice in two-dimensional space. Am. Political Sci. Rev. **95**(1), 155–167 (2001)

Laver, M., Shepsle, K.: Making and Breaking Government: Cabinets and Legislatures in Parliamentary Democracies. Cambrige University Press, New York (1996)

McKelvey, R.D.: General conditions for global intransitivities in formal voting models. Econometrica **47**, 1085–1112 (1979)

McKelvey, R.D.: Covering, dominance, and institution-free properies of social choice. Am. J. Political Sci. **24**, 68–96 (1986)

McKelvey, R.D., Schofield, N.: Structural instability of the core. J. Math. Econ. **15**(3), 179–198 (1986)

McKelvey, R.D.: Intransitives in multidimensional voting models and some implications for agenda control. J. Econ. Theory **12**(3), 472–482 (1976)

Miller, N.R.: A new solution set for tournaments and majority voting. Am. J. Political Sci. **24**, 68–96 (1980)

Miller, N.R.: In search of the uncovered set. Political Anal. **15**, 21–45 (2007)

Mordeson, J.N., Clark, T.D.: The existence of a majority rule maximal set in arbitrary n-dimensional spatial models. New Math. Nat. Comput. **6**, 261–274 (2010)

Mordeson, J.N., Clark, T.D.: Representing thick indifference in spatial models. Adv. Fuzzy Sets Syst. **12**, 79–91 (2012)

Mordeson, J.N., Clark, T.D., Gibilisco, M.B., Casey, P.C.: A consideration of different definitions of fuzzy covering relations. New Math. Nat. Comput. **6**, 247–259 (2010)

Mordeson, J.N., Clark, T.D., Miller, N.R., Casey, P.C., Gibilisco, M.B.: The uncovered set and indifference in spatial models: a fuzzy set approach. Fuzzy Sets Syst. **168**, 89–101 (2011)

Mordeson, J.N., Nair, P.: Successor and source functions of (fuzzy) finite state machines and (fuzzy) directed graphs. Inf. Sci. **95**, 113–124 (1996)

Penn, E.M.: Alternative definitions of the uncovered set, and their implications. Soc. Choice Welf. **27**(1), 83–87 (2006)

Plott, C.R.: A notion of equilibrium and its possibility under majoirty rule. Am. Econ. Rev. **57**(4), 787–806 (1967)

Richelson, J.: A comparative analysis of social choice functions, iv. Behav. Sci. **35**, 346–353 (1981)

Richelson, J.: Social choice solution sets. Analytical Assessments Corporation (1980)

Riker, W.H.: Implications from disequilibrium of majority rule for the study of institutions. Am. Political Sci. Rev. **74**, 432–446 (1980)

Schofield, N.: Existence of a structurally stable equilibrium for a non-collegial voting rule. Public Choice **51**(3), 267–284 (1986)

Shepsle, K.: Institutional arrangements and equilibrium in multidimensional voting models. Am. J. Political Sci. **23**(1), 27–59 (1979)

Shepsle, K., Weingast, B.: Uncovered sets and sophisticated voting outcomes with implications for agenda institutions. Am. J. Political Sci. **31**, 49–72 (1987)

Sloss, J.: Stable outcomes in majorty rule voting games. Public Choice **15**(1), 19–48 (1973)

Strom, K., Laver, M.J., Budge, I.: Constraints on cabinet formation in parliamentary democracies. Am. J. Political Sci. **38**(2), 303–335 (1994)

Theis, M.F.: Keeping tabs on partners: the logic of delegation in coalition governments. Am. J. Political Sci. **45**(3), 580–598 (2001)

Tovey, C.A.: The instability of instability. Technical report, NPSOR-91-15. Monterey, CA: Department of Operations Research, Naval Postgraduate School (1991)

Warwick, P.V.: Ministerial autonomy or ministerial accommodation? constested bases of government survival in paliamentary democracies. Br. J. Political Sci. **29**(2), 369–394 (1999)

Chapter 6
Predicting the Outcome of the Government Formation Process: A Fuzzy Two-Dimensional Public Choice Model

Abstract Under the most basic of assumptions of Euclidean preferences, majority rule erupts into cycling in two or more dimensional space, and no alternative remains undefeated. The resulting McKelvey's Chaos Theorem (McKelvey 1976) forces scholars to reconsider basic assumptions about the rational behavior of political actors and their attempts to form coalitions. The government formation literature remains divided on how to best solve the problem. More recently, proposed models either assume cabinet ministers are virtual dictators over their policy jurisdiction (Laver and Shepsle 1996) or rely on complex game-theoretic arguments, which do not lend themselves to empirical verification (Baron 1991; Diermeier and Merlo 2000). This chapter builds on the fuzzy maximal set model developed in Chap. 4. It presents a fuzzy maximal set multi-dimensional model to predict the outcome of the government formation process. We conclude by comparing the predictions made by the model using CMP against actual governments formed after European Parliamentary elections between 1945–2002.

6.1 A Two-Dimensional Fuzzy Model

Conventional government formation models in multi-dimensional policy space contain the same assumptions (and restrictions) as those of the single dimensional models. However, in multi-dimensional policy space, these models break down owing to preference cycling (McKelvy's Chaos Theory (McKelvey 1976)). McKelvey's theorem argues that under majority rule in minimally two-dimensional space, cycling occurs resulting in the ability for any alternative in policy space to be chosen, even if another alternative would benefit every actor. The government formation literature offers a variety of models with additional restrictions to attempt to deal with this problem. Assuming parties are solely office-seeking, Riker's (1962) model results in predicting only minimum winning coalitions. Other models assume bargaining occurs (Grofman 1982) or that parties strategically place ministers to maximize the

P. C. Casey et al., *Fuzzy Social Choice Models*, Studies in Fuzziness
and Soft Computing 318, DOI: 10.1007/978-3-319-08248-6_6,
© Springer International Publishing Switzerland 2014

likelihood of policy control on certain issues (Laver and Shepsle 1996). Yet others, assume a state of policy equilibrium as an exception to the norm (Schofield 1993), or reject traditional models completely in favor of game-theoretic models lacking extensive empirical verification (Baron 1991; Diermeier and Merlo 2000).

In the previous chapter, we presented a formal argument showing that the application of fuzzy preferences and the fuzzy maximal set offers a simple approach to addressing the problem of chaos in multi-dimensional public choice models. In this chapter we construct and test the ability of a fuzzy public choice model to predict outcomes of the government formation process.

The fuzzy multi-dimensional model incorporates thick indifference to represent players' preferences, and it makes use of the fuzzy maximal set to predict outcomes. As we will demonstrate, a fuzzy model also increases the ability of public choice models to represent non-separable dimensions of policy space.

We test the fuzzy multi-dimensional model using the same CMP data that we used to test the single-dimensional model. Once again using the Benoit et al. (2009) bootstrapping approach, we resample the data to increase the number of positional codings by 1,000 for each party on each policy dimension. We then identify finance and foreign affairs as the two policy dimensions that are the most salient for parties as they approach the task of forming a government. These two policy dimensions are combined and a two-dimensional kernel density estimation is used (Wand and Jones 1995). This density estimation is scaled between 0 and 1 to reflect a party's preference. After this process, Definition 6.5 scores the possible coalitions into the $0, 0.25, 0.5, 0.75, 1$ interval. This designates the level of maximality for a given coalition. Coalitions registering the highest scores, or in the highest maximal level, are assumed to be more likely to form a coalition. The predictions of the multi-dimensional fuzzy government formation model are then compared with the actual governments that formed.

6.2 A Fuzzy Model of Government Formation

Models of government formation that include thick indifference in actors' preferences allow researchers to escape the chaos resulting from intransitivity in social preference. However, previous public choice models incorporating indifference have been overly complicated and do not lend themselves to empirical verification. As we showed in the previous chapter, fuzzy set theory provides a simpler method for modeling indifference. Moreover, a fuzzy public choice model greatly reduces the problems introduced by intransitivity without imposing exogenous restrictions. Here we present and test a multi-dimensional public choice model to predict the outcome of the government formation process in parliamentary democracies. The model relies on a procedure using the fuzzy maximal set which allows parliamentary coalitions to be understood as more or less maximal instead of defeated or undefeated.

Fuzzy set theory differs from traditional set theory because it asks to what degree is an element in a set instead of is an element in or out of a set. While the set inclusion

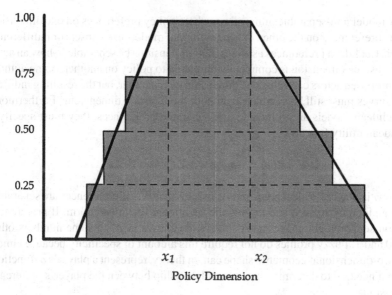

Fig. 6.1 Discrete fuzzy preference profile

level in traditional set theory must be 1 or 0, either in or out of the set, fuzzy set theory uses a characteristic function, which we call σ_i, to map the universe X into the interval [0, 1], formally written as $\sigma_i : X \rightarrow [0, 1], \forall i \, \varepsilon \, N$. In policy terms, $\sigma_i(x)$ refers to the degree to which x is an ideal policy of individual i, where 1 is an ideal policy and 0 is a completely non-ideal policy.

As the reader will recall, in chapter two we discussed several ways to conceptualize σ_i over a along a single policy dimension. We ultimately settled upon a discretized fuzzy preference, which we incorporated into our one-dimensional model in chapter four. In that representation of a player's preference, if σ is restricted to a discrete set, then a player is said to possess discretized preferences, or "thick- indifference."[1] Figure 6.1 presents such a profile. In this example, σ_i is an element of 0, 0.25, 0.5, 0.75, 1. An actor is completely indifferent over all alternatives at a specific preference level because all alternatives have an equivalent σ_i. While researchers can choose any numbers of preference levels, a Likert-esque division has a natural appeal. A preference level of 1 can be interpreted as perfectly ideal, a level of 0.75 as almost ideal, a level of 0.5 as neither ideal nor not ideal, a level of 0.25 as not ideal and a level of 0 as completely not ideal.

The same logic applies to two-dimensional models. In a two-dimensional fuzzy profile, σ_i becomes a third dimension perpendicular to both the X and Y policy dimensions. Adding a second dimension allows the model to break even further away from assumptions imposed by the conventional approach. Fuzzy set theory can

---

[1] Formally, an actor i possesses thick indifference if for $a, b \in X$, $a \neq b \not\Rightarrow aP_i b$ or $bP_i a$ where P_i is the strict preference relation for player i.

easily model non-separable dimensions, where policy preferences on one dimension affects preferences on the other. When traditional models use winsets or indifference curves, Euclidean preferences assumes their dimensions be separable —how an actor perceives one dimension is completely unrelated to policy on another. Some authors have proposed actors can value one dimension over another, but the resulting indifference curves must still be symmetric on both the X and Y dimension.[2] Furthermore, if Euclidean models are to include non-separable preferences, they must specify a Euclidean utility function,

$$u(y) = -(y - x_i)A_i(y - x_i)^T,$$

where x_i is player i's ideal point and A_i is 2×2 matrix.[3] If preferences are separable, A_i is an identity matrix, and $u(y)$ is the traditional Euclidean norm. If preferences are non-separable the model must specify the entries in A_i to some numbers other than 0 and 1. Fuzzy profiles do not require this amount of specificity because almost any two-dimensional geometric shape can, in theory, represent a player's σ-function, without needing to determine the exact relationship between the player's preference dimensions.

Figure 6.2a, b illustrate the differences between separable and non-separable dimensions. Figure 6.2a shows separable indifference curves, which are symmetric in relation to the x and y axes. The circle signifies an actor values both the X and Y dimensions equally while the ovals signify greater concern for policy on one dimension. These are separable because the symmetry implies an actor has a solitary ideal point on both the X and the Y dimensions, which does not vary given the policy position on the opposing dimension. Figure 6.2b shows non-separable indifference curves, which are not symmetric in relation to the x and y axes, implying policy preferences on one dimension are related to preferences on the other.

In addition to modeling non-separable issue dimensions, fuzzy preference profiles move beyond other assumptions about indifference curves. First, fuzzy profiles do not require convex or pseudo-convex indifference curves. The set-theoretic approach frees researchers from using a series of differentiable equations to calculate areas of indifference, making the modeling process less cumbersome and allowing for concave preferences. Second, fuzzy profiles can model actors with two cores, or two ideal regions. An actor can have two cores when σ_i is a multimodal function. Figure 6.2c displays a fuzzy, concave, multimodal profile. While it is beyond the scope of this chapter to debate whether an actor, either an individual or a group of individuals, can have two ideal points in policy space, fuzzy spatial models have the flexibility to include such possibilities where one party possess two or more distinct legislative factions. Subsequently, we do not place any restrictions on indifference curves in my analysis.

Once there is a process to construct fuzzy preference profiles for every actor, predicting coalition formation then becomes an analysis of intersecting indifference

---

[2] See Laver and Hunt (1992) and Laver and Shepsle (1996).

[3] In k-dimensional Euclidean space, A_i would be a $k \times k$ matrix.

Fig. 6.2 Difference between traditional and fuzzy indifference curves. **a** Seperable. **b** Seperable. **c** Multimodal, concave and fuzzy

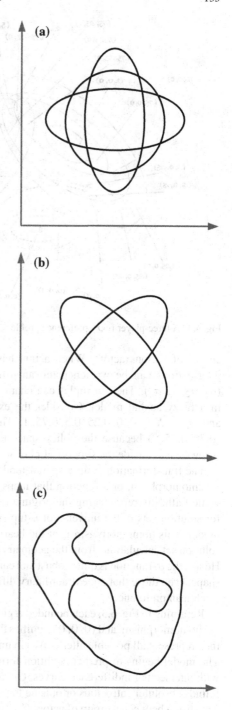

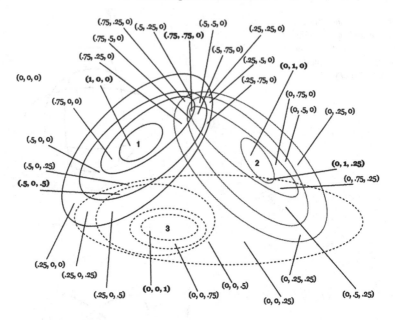

Fig. 6.3 A three-player fuzzy preference profile

curves of various actors. When actors have intersecting indifference curves, the intersections can be written using an n-tuple with corresponding σ_i written as $(\sigma_1, \sigma_2, ..., \sigma_n)$. These n-tuples can characterize all possible discrete policy areas in a fuzzy spatial model. Consider the example in Fig. 6.3, where $N = 1, 2, 3$ and $\sigma_i : X \rightarrow 0, 0.25, 0.5, 0.75, 1$. The intersection labeled w is written as $(0.75, 0.75, 0)$ because the policy space lies within player 1 and player 2's 0.75 α-levels and outside the support of player 3.

The transformation from a spatial model into a universe of n-tuples represents a homomorphism, or a function that maps a relation from one set into another set while faithfully reproducing the original relation. As a consequence, we can solve for solution sets in the universe of n-tuples and apply these findings to the spatial model. This homomorphism is at the heart of fuzzy spatial model because it frees solution set calculations from the geometry and mathematics of indifference curves. Hence, the σ-function is quite robust and can be represented by almost any geometric shape, even those that are extraordinary difficulty, if not impossible, to capture with Euclidean preferences.

Returning to Fig. 6.3, each bounded region can be transformed in an n-tuple using the homomorphism, and $(0, 0, 0)$ signifies the area absent of all profiles. The n-tuples then represent all possible alternatives in universe X coalitions could possibly form. The model begins to predict coalition formation by analyzing the group of actors with intersecting indifference curves. Obviously, an actor will not agree to a policy it finds absolutely atrocious or outside the support of its fuzzy profile. Accordingly, a coalition between a group of actors $C = \{1, 2, ..., c\}$ is possible if and only if there exists an n-tuple in X where $\sigma_i > 0$ for all c in C.

Just because a coalition is possible does not mean that it will form a government. Hence, it becomes necessary to add other formation criteria to the model. The most common and obvious criterion in the government formation literature is majority support. If a government does not have the support of the legislature it cannot survive. While this logic may seemingly exclude minority parties, set theory allows the model to predict these types of government. For instance, if the two parties in a possible coalition $C_M = 1, 2$ control more than 50 % of the legislatures in parliament, either party can form a minority government as long as it keeps government policy within the support of the other's preferences. Applying this method, the power set of any possible majority coalition C, denoted $P(C)$, is the prediction set, where some coalition $S \subset P(C)$ does not need to contain a majority of the legislatures itself.

Because $|P(C)| = 2^c$, where c is the number of parties in C, it is easy to see that a majority criterion does not effectively reduce the size of a government formation prediction. Reconsider the example in Fig. 6.3 and assume each fuzzy profile represents a party in parliament with one-third of the legislators. Under these assumptions, the prediction set consist of $(\{1\}, \{2\}, \{3\}, \{1, 2\}, \{1, 3\}, \{2, 3\})$ because every coalition is possible besides the grand coalition, $\{1, 2, 3\}$.[4] If C included five, six, or even more parties, the number of predicted coalitions increases quickly. It then becomes necessary to add another criterion to limit the size of the prediction set.

One method proposed considers majority supported, Pareto efficient alternatives when determining possible coalitions (Casey et al. 2012).

Definition 6.1 (*The Pareto Set*) The Pareto set is defined by Austen-Smith and Banks (1999) as follows, where ρ is a preference profile and $P(a, b)$ means a is strictly preferred to b:

$$PS_L(\rho) = \{x \in X \mid \forall y \neq x, \ P(y, x; \rho) \cap L \neq \emptyset \implies P(x, y; \rho) \cap L \neq \emptyset\}$$

In words, if an alternative x is Pareto efficient, then any member of coalition L preferring some other alternative y to x implies another member of L strictly prefers x to y. The Pareto set of universe X is then the union of all $PS_L(\rho)$ for every legislative coalition $L \subseteq N$. The bolded labels in Fig. 6.3 illustrate the Pareto set. More simply, a Pareto efficient alternative is such that a move away from it implies that at least one player is made worse off even though some other player will benefit.

When applying Pareto efficiency together with the majority-supported criterion, the prediction set does not change. This occurs because if an alternative x represents a possible coalition C, x is either a Pareto efficient alternative or there exists an alternative w such that for one member of the coalition $\sigma_i(w) > \sigma_i(x)$. In the latter case, w would be Pareto efficient and the original coalition is still predicted or there is another alternative that is greater than w. However, while Pareto efficiency suggest what policy areas are ultimately accepted by a particular coalition, it does not constrain the set of possible coalitions.

---

[4] The reader should note there does not exist any n-tuple in Fig. 6.3 where $\sigma_i > 0$ for all three players.

Another method common in the literature is to identify undefeated alternatives, or in other words alternatives with an empty winset. Undefeated alternatives make up the maximal set.

Definition 6.2 (*The Maximal Set*) The maximal set is formally defined by Austen-Smith and Banks (1999), where R is the weak preference relation:

$$M(R, X) = \{x \in X \mid \forall y \in S \;\; xRy\}$$

Or, the maximal set contains all alternatives in X that are as least of good as all the other alternatives in X. More simply, the maximal set is the set of undefeated alternatives. In two dimensions, the maximal set is almost always empty under Euclidean preferences.[5] Fuzzy preferences greatly reduces the likelihood of cycling (Clark et al. 2008); however, cycles still exist with fuzzy preferences. Returning to the example in Fig. 6.3, $M(R, X) = \{\emptyset\}$ because every alternative is defeated.[6] There does not exist an undefeated alternative, so another solution set must be used.

Not only is there a practical problem with using the maximal set as the prediction set, there is also a theoretical problem. The maximal set requires a set decision rule to determine which alternatives are undefeated, but no such rule may exist during coalition bargaining. Furthermore, even when applied to fuzzy preferences, it requires an actor to choose x or y or be indifferent between the two, without allowing for greater description. For example, given two alternatives x and y and given $\sigma_i(x) > \sigma_i(y)$, individual i will always prefer x to y to the same degree regardless of the difference between $\sigma_i(x)$ and $\sigma_i(y)$. More specifically, assuming $\sigma_i(y) = 0$, individual i will always prefer x to y to the same degree even if $\sigma_i(x) = 1$ or $\sigma_i(x) = 0.001$. This is the dichotomy inherent in traditional preferences. Thus, while the goal of this chapter is propose fuzzy preference profiles with thick indifference, the use of the maximal set would restrict the model to traditional preferences, where xPy or xIy.

A fuzzy preference relation allows researchers to model this type of behavior, where an actor may prefer an alternative to another but is still indifferent between the two, to a degree. Definition 6.3 proposes such a relation ρ, where $\wedge$ represents the minimum function and r is an individual risk factor. $\rho_1(x, y)$ allows an actor to prefer x to y and y to x with varying level of strength.

Definition 6.3 (*Individual Preference Relation*)

$$\rho_i(x, y) = \begin{cases} [\sigma_i(x) - \sigma_i(y) + r] \wedge 1 & \text{if } \sigma_i(x) \geq \sigma_i(y) \\ 1 - [(\sigma_i(y) - \sigma_i(x) + 1 - r) \wedge 1] & \text{if } \sigma_i(x) < \sigma_i(y), \end{cases}$$

where $r \in [0, 1]$.

---

[5] See Schofield (1993) for an important exception.

[6] When the maximal set is empty under thick indifference, Mordeson et al. (2011) propose using a fuzzy uncovered set. Nonetheless, the fuzzy uncovered set is a less than ideal solution set because it is almost always the Pareto set, thus not effectively reducing the number of predicted coalitions.

Definition 6.3 has several desirable qualities. First, the variable r accounts for the risk an actor is willing to accept. If $r = 0$, individual i will prefer one alternative to another less strongly than when $r = 1$. For example, if $r = 0$, $\sigma_1(x) = 0.75$ and $\sigma_1(y) = 0.25$, $\rho_1(x, y) = 0.5$ and $\rho_1(y, x) = 0$. In contrast, if $r = 1$, $\rho_1(x, y) = 1$ and $\rho_1(y, x) = 0.5$.[7] Second, $\rho_1(x, y)$ maps into the interval $[0, 1]$, which is an important criteria for a fuzzy preference relation. Third, $\rho_1(x, y)$ and $\rho_1 i(y, x)$ reproduces the inequality relationship between $\sigma_1(x)$ and $\sigma_1(y)$.[8] In addition, when $r = 1$, $\rho_1(x, y)$ is reflexive and complete.[9] Because reflexivity and completeness are important qualities in normal preference literature and because they are required for certain choice functions to be considered rationalizable (Georgescu 2007), it is assumed $r = 1$ in future reference. The simplified version of $\rho_i(x, y)$, where $r = 1$, is defined in Definition 6.4

Definition 6.4

$$\rho_i(x, y) = \begin{cases} 1 & \text{if } \sigma_i(x) \geq \sigma_i(y) \\ 1 - \sigma_i(y) + \sigma_i(x) & \text{if } \sigma_i(x) < \sigma_i(y) \end{cases}$$

The use of traditional preferences in maximal set calculations makes the solution less than ideal for government formation because there is no guarantee there will exist an undefeated alternative. However, the fuzzy preference relation $\rho_i(x, y)$ applies to other solutions sets that do not require traditional preferences. The fuzzy maximal set, proposed by Dasgupta and Deb (1991) and later Georgescu (2007), uses a fuzzy characteristic function and preference relation to infer the degree an alternative is maximal for an individual i. Definition 6.5 defines the fuzzy maximal set, where $\vee$ is the maximum function and $\text{supp}(\sigma_1)$ refers to the set of alternatives where $\sigma_1 > 0$. In words, the fuzzy maximal set answers to what degree is an alternative x a maximal element considering its characteristic σ_1 and its relationship with other alternatives in universe X.

Definition 6.5 (*Individual Maximal Set*)

$$M_{G_i}(\rho_i, \sigma_i)(x) = \sigma_i(x) \wedge$$
$$\bigwedge \left\{ \bigvee \{t \in [0, 1] \mid \sigma_i(y) \wedge \rho_i(y, x) \wedge t \leq \rho_i(x, y)\} \mid y \in \text{supp}(\sigma_i) \right\}$$

As it stands, $M_{G_1}(\rho_1, \sigma_1)$ does not predict a specific coalition. The definition does, however, provide a method to aggregate σ-functions and a fuzzy preference relation to create a relational maximal set. $M_{G_1}(\rho_1, \sigma_1)$ has several desirable qualities. For instance, $\sigma_1(x) = 1$ implies $M_{G_1}(\rho_1, \sigma_1)(x) = 1$, meaning an actor finds an

---

[7] $\rho_i(x, y) = (0.75 - 0.25 + r) \wedge 1$ because $\sigma_i(x) \geq \sigma_i(y)$. $\rho_i(y, x) = 1 - [(0.75 - 0.25 + 1 - r) \wedge 1] = 1 - [(0.5 + 1 - r) \wedge 1$ because $\sigma_i(y) \geq \sigma_i(x)$.

[8] See Appendix B for the formal argument.

[9] See Appendix B for the formal argument.

ideal element completely maximal in X.[10] In addition, the size of $\sigma_1(x)$ constrains $M_{G_1}(\rho_1, \sigma_1)(x)$. Thus, $M_{G_1}(\rho_1, \sigma_1)(x)$ can be no greater than $\sigma_1(x)$.[11] This constraint is necessary; an alternative is maximal to the extent that an actor views it as an ideal policy. Furthermore, when $r = 1$, $M_{G_1}(\rho_1, \sigma_1)(x) = \sigma_1(x)$.[12] The equality has an intuitiveness about it, where an actor's preferences are directly related to the actor's maximal set calculations over the alternatives.

Because Definition 6.5 is focused on the individual, a framework which uses the maximal set to predict coalitions must be constructed. More specifically, there must be a function to aggregate individual maximal sets based on possible coalitions. In fuzzy mathematics, t-norms are often used as aggregators. T-norms are binary operators used to measure logical intersection or truthfulness.[13] For example, a common t-norm is the Gödel t-norm, or more commonly referred to as the minimum function, which was used in Definitions 6.3 and 6.4. For example, two fuzzy numbers characterizing some attribute, let's say 0.5 and 0.8 have a logical intersection at $\min(0.5, 0.8) = 0.5$ because both numbers represent at least the 0.5 level of the attribute. When applying this logic to coalitions, a coalition C is maximal at the minimum of $M_{G_1}(\rho_1, \sigma_1)(x)$ values for all $c \in C$. In other words, the viability of coalition C is determined by the most displeased coalition member. This rejects a type of consensus between the coalition members. Once individual $M_{G_1}(\rho_1, \sigma_1)$ levels are determined for each coalition at each alternative in X, the maximum level of possibility for a coalition is determined by taking the maximum across all alternatives which represent intersections of the specific coalition. This final step selects the alternative with the highest level of cooperation from all members in the coalition. Definition 6.6 formally defines this procedure.

Definition 6.6 Let $C = (1, 2, ..., c\} \subseteq N$ and $x_1, x_2, ..., x_k \in X$.

$$M_G(C) = \bigvee \Big(\bigwedge \big(M_{G_1}(\rho_1, \sigma_1)(x_1), M_{G_2}(p_2, \sigma_2)(x_1), \ldots, M_{G_c}(\rho_c, \sigma_c)(x_1)\big),$$
$$\bigwedge \big(M_{G_1}(\rho_1, \sigma_1)(x_2), M_{G_2}(p_2, \sigma_2)(x_2), \ldots, M_{G_c}(\rho_c, \sigma_c)(x_2)\big),$$
$$\vdots$$
$$\bigwedge \big(M_{G_1}(\rho_1, \sigma_1)(x_k), M_{G_2}(p_2, \sigma_2)(x_k), \ldots, M_{G_c}(\rho_c, \sigma_c)(x_k)\big)\Big).$$

Furthermore, Definition 6.7 simplifies the procedure, taking into account that when $r = 1$, $M_{G_i}(\rho_1, \sigma_1)(x) = \sigma_1(x)$.

---

[10] See Appendix B for the formal argument.

[11] See Appendix B for the formal argument.

[12] See Appendix B for the formal argument.

[13] For a more thorough discussion on t-norms see *Triangular Norms* by Klement and Pap (2000).

Table 6.1 Maximal set calculations for pareto elements

Alternative	$M_{G_1}(\rho_1, \sigma_1)$	$M_{G_2}(\rho_2, \sigma_2)$	$M_{G_3}(\rho_3, \sigma_3)$
(0.75, 0.75, 0)	0.75	0.75	0
(0, 1, 0.25)	0	1	0.25
(0.5, 0, 0.5)	0.5	0	0.5

Table 6.2 Maximal coalition calculation

Coalition	(0.75, 0.75, 0)	(0, 1, 0.25)	(0.5, 0, 0.5)	$M_G(C)$
{1,2}	min(0.75, 0.75) = 0.75	min(0, 1) = 0	min(0.5, 0) = 0	0.75
{1,3}	min(0.75, 0) = 0	min(0, 0.25) = 0	min(0.5, 0.5) = 0.5	0.5
{2,3}	min(0.75, 0) = 0	min(1, 0.25) = 0.25	min(0, 0.5) = 0	0.25

Definition 6.7 Let $C = \{1, 2, ..., c\} \subseteq N$.

$$M_G(C) = \bigvee \Big(\bigwedge (\sigma_1(x_1), \sigma_2)(x_1), ..., \sigma_c(x_1)) \, ,$$
$$\bigwedge (\sigma_1(x_2), \sigma_2(x_2), ..., \sigma_c(x_2)) \, ,$$
$$\vdots$$
$$\bigwedge (\sigma_1(x_k), \sigma_2(x_k), ..., \sigma_c(x_k)) \Big)$$

These longer, denser equations are useful in better understanding the mechanics behind the aggregation method. A further condensed equation is offered below.

Definition 6.8 (*Aggregated Fuzzy Maximal Set*)

$$M_G(C) = \bigvee_{i=1}^{k} \bigwedge_{c \in C} M_{G_c}(\rho_c, \sigma_c)(x_i)$$

Returning to the example in Fig. 6.3, Definition 6.8 can be applied to produce a more distinct prediction set. The first step is to determine the majority supported, possible coalitions. The possible coalitions are represent by the set $\{\{1, 2\}, \{1, 3\}, \{2, 3\}\}$. Since there is no 3-tuple where all $\sigma_1 > 0$, $\{1, 2, 3\}$ is excluded. The second step is to calculate fuzzy maximal set values for alternatives in X. Table 6.1 reports the maximal set values for the three Pareto efficient alternatives instead of all the alternatives for the sake of simplicity, without affecting the final outcome. Once individual maximal sets are calculated, the next step is to take the minimum value for each coalition member on each alternative. These results are reported in the first four columns of Table 6.2. The final step is to take the maximum across player to obtain the final value for $M_G(C)$. $M_{G_I}(C)$ tells us which coalition is more likely to form.

In this example, the most viable coalition would be $\{1, 2\}$ at a 0.75 maximal set level. The least likely possible coalition to form would be 2, 3 at a 0.25 level. The predicted coalition will be a subset of 1, 2, which includes minority governments of 1 and 2. This solves the possible over-prediction problem of other methods where too many coalitions may be included in the final prediction set, while retaining the possibility of minority governments. $M_{G_i}(C)$ essentially creates a hierarchal structure to the coalition prediction set. While it is quite possible for more than one coalition to be maximal at a certain level, $M_G(C)$ does create a more limited and nuanced prediction set.

Applied to spatial policy preferences, fuzzy set theory characterizes an infinite policy space into discrete policy spaces based on the degree to which a range is included in an actor's ideal policy range. However, fuzzy preferences alone may not produce a satisfactorily small prediction set. Using the fuzzy maximal set together with the fuzzy preference relation presented in this section, an actor's maximal policy choice is identical to the actor's most ideal policy area. This relationship may seem redundant, but it emphasizes the connection between individual preferences and individual choice in the model. Most importantly, the model presented here aims to capture the consensus needed for coalition formation. Regardless of size, all parties must agree when entering into coalition, and each possesses the power to end the specific coalition at any given time. The model's logic rejects this aspect of coalition formation by requiring the member with the smallest preference for the coalition to define to overall viability of the coalition itself. Any coalition is only as strong as its weakest member.

6.3 Fuzzifying the CMP

As has been consistent throughout this work, we once again employ the exhaustive Comparative Manifesto Project's database for party preference extraction. The decision to use manifesto codings instead of role-call votes is appealing because it does not make the analysis susceptible to tautologies because it is difficult to ascertain whether coalitions are a function of party votes or party votes are a function of coalitions (Laver 2001). In addition, the CMP data set is quite expansive and allows us to test the model on a potential 3,108 elections in 54 countries without having to survey experts or collect raw manifestos. The data comprise manifestos for each party in each election and score every quasi-sentence, similar to grammatical clauses, in party manifestos on one of 57 policy dimensions (referred to as PER categories), including an uncoded category.

The CMP coding scheme takes naturally ambiguous party statements about military force, privatization, social justice, and other policy areas and quantifies them into exact locations on a number line. Obviously, this does not conform to the concept of fuzzy spatial preferences presented in this chapter. Thus, the task becomes three-fold when restoring indifference to policy positions, which have been boiled down into a minute point through human coding: (1) the data needs to be resampled to estimate the indifference in party preferences, (2) relevant policy dimensions must be chosen or constructed from the CMP categories, and (3) fuzzy preferences need to be constructed for each party from the outputs of tasks one and two.

The first task involves fuzzifying party positions in the CMP. Benoit, Laver, and Mikhaylov's (2009) bootstrapping procedure is highlighted in chapter four. As the reader will recall, the data are resampled to reproduce an additional 1,000 codings for every political party entry in the CMP. The resamples are based on a multinomial distribution where the codings in each PER categories are the given probabilities for the generated data. The resulting 1,000 codings are similar to the original manifesto; however, some PER categories may lose a percentage of quasi-sentences while others may gain, creating another possible policy position on each dimension. The total number of quasi-sentences remain constant in each manifesto. The result is an estimation of variance in each party's position on every PER dimension that declines as the length of a particular manifesto increases. The PER categories remain additive, and now there is an estimation variance in party positions.

This approach argues that political parties are indifferent over sizable policy areas rather than being unable to explain their party platform. This indifference is then responsible for the variance that emerges from the bootstrapping procedure. Error in party position is not noise, but rather it is politicians expressing indifference over a policy space. The subsequent results of the bootstrapping procedure provide other possible party positions, which have a central tendency and an ultimate range. This newly derived variance is necessary in the construction of fuzzy preferences.

The second task is to develop policy dimensions that represent the dimensions relevant in coalition bargaining in European democracies. While issue dimensions during legislative elections may vary from country to country (Budge et al. 1987), existing evidence suggests issue dimensions in government formation remain largely stable. Several studies have shown success using either a left-right dimension or other dimensions used across countries and elections (Laver and Budge 1993; Laver and Hunt 1992; Laver and Schofield 1990; Brown and Feste 1975). While some models utilize a less general approach (Laver and Shepsle 1996), deriving some underlying issue dimensions from the CMP data is quite complicated due to the fact the seemingly related variables such as positive and negative military references are uncorrelated (Van der Brug 2001; Elff 2009). Because common statistical techniques relying on correlation are unable to detect underlying factors in the data, Van der Brug (2001) suggests using multidimensional scaling to estimate party positions. The use of this technique may be unwarranted, however, because expert surveys show key portfolios in ministry allocation are quite similar across countries (Laver and Hunt 1992). The finance and foreign affairs portfolios are almost always considered the most salient. One explanation for the importance of these portfolios could be parties are the most eager to influence these types of policies. Another explanation is that regardless of other contentions, governments must address the economy and foreign relations almost immediately when taking office. Following this reasoning,

Table 6.3 Descriptive statistics of CMP

Dimension	Mean	Median	Minimum	Maximum
Economic	1.26	0	−74.30 (1976 Luxembourg communist party)	63.80 (1973 Norway people's party)
Foreign relations	3.05	1.74	33.40 (1960 Denmark socialist people's party)	57.14 (1988 Denmark conservative people party)

the model uses a priori policy dimensions on economic and foreign affairs policy. Laver and Budge (1992) create several additive dimensions using the CMP's PER categories, which are used throughout the literature concerning the formal modeling of political parties. The model uses their planned and market economic dimensions to create a broader economic dimension and the external relations PER categories selected for the *rile* dimension to construct a foreign affairs dimension.[14] Table 6.3 provides the descriptive statistics of the original CMP positions.

Once the economic and foreign relations dimensions are combined in two-dimensional space, probability density functions for each party can be estimated using the 1,001 probable party positions. Kernel density estimation is a nonparametric procedure used to estimate probability density functions of a random variable using kernels or bins from the variables histogram (Wand and Jones 1995).[15] Once the density of positions have been estimated, the density is scaled from 0 to 1. The resulting three-dimensional curve resembles the preference curves presented in the previous section if the perspective is perpendicular to the z-axis representing the density of party positions. Because the density is scaled between 0 and 1, the preferences can be contoured at the predetermined alpha levels. Thus, the placements of the bootstrapped party positions are intrinsically linked to fuzzy party preferences, with the density function being the measure of party idealness, σ_1.

Once the preferences are extracted from each party, we apply the fuzzy maximal set Definition 6.8 described in the previous section over all the possible coalitions. The definition scores each coalition between 0 and 1 based on the degree to which each coalition is maximal. The predicted coalitions were then compared with real-world governments in European democracies.

---

[14] A party's economic policy is constructed by the following formula: per414 + per401 − per412 − per404 − per403. A party's foreign affairs' policy is constructed by the following formula: per104 − per107 − per106 − per105 − per103.

[15] Because the bootstrapped CMP comprises discrete policy dimensions for each party, this book uses a specific design for such purposes. For further information on two-dimensional kernel density estimation see Wand and Jones (1995). In addition, for more explanation on the interaction Kernel estimation and bootstrapping, see Schucany and Polansky (1997).

6.4 Empirical Performance of the Fuzzy Model

Fuzzy preferences were constructed for 97 European parliamentary elections out of a possible 235 elections. The number of elections was limited due to complications with the bootstrapped data. If a party had a zero additive score on either its economic or foreign policy, the bootstrap procedure was unable to estimate the variance in party position. Subsequently, any election with one of these parties was dropped from the analysis. While the results presented in this chapter are necessarily preliminary, the relatively good fit between the predicted and actual government give reason for confidence in the consensus model developed.

Table 6.4 summarizes the performance of the fuzzy formation model divided among the possible types of parliamentary governments.[16] Of the governments tested, the fuzzy model correctly predicts 56.7 % (55) of the cases. Besides the trivial case of single-majority governments, of which 100 % were correctly predicted, the best prediction rate is for minority governments, of which 72.2 % were correctly predicted. The next best prediction rate is for minority coalition governments, of which 62.5 % were correctly predicted, followed by supermajorities, of which 43.75 % of cases were correctly predicted. The least accurate predictions are for minimum-winning coalitions, which still provided a respectable 43.1 % prediction rate.

These first results represent all types of possible coalitions, where party preferences must merely intersect for the model to predict a government. However, because of the procedure presented in this chapter, the fuzzy maximal set ranks each coalition by mapping possible coalitions into the 0, 0.25, 0.5, 0.75, 1 discrete interval. This ranking determines which coalitions are more maximal than others in each parliamentary election. While there may not exist a coalition with a maximal score of 1 in each election, there will be least two levels of coalitions, at zero and at some other level in the discrete interval. Table 6.4 demonstrates the ability of fuzzy maximal set to identify the actual government as the most maximal coalition. The first-level maximal column in Table 6.4 displays the percentage of the cases where the largest maximal coalition correctly predicts the government formed in the real-world. Similarly, the other-level maximal column displays the percentage of cases where the largest maximal coalition did not correctly predict the actual government.

The highest maximal score correctly predicts about two-fifths of the cases (44.3 %). It most accurately predicts single-party minority governments, followed by minimum-winning coalition governments. Other maximal scores correctly predict 4.1 % of cases. They most accurately predict minimal-winning coalition governments at 12.5 %. On average, the maximal set procedure generated between two and three levels (mean = 2.8) of maximal coalition scores. One significant problem with the analysis occurs when the model fails to make a prediction. This happens because elections do not contain sufficient overlap in party preferences. In roughly 25 % of cases, the intersection of parties' indifference curves do not produce some type of majority coalition. Removing these cases greatly increases the accuracy of the model.

---

[16] Percentages rounded to nearest tenth.

Table 6.4 Summary of results for 2-D analysis by government type

Type (Number)	1st-level maximal (%)	2nd-level maximal (%)	Other level maximal (%)	No prediction (%)	% correct	Average # levels
Single-party minority (18)	12 (66.67)	1 (5.56)	0 (0.00)	4 (22.22)	13 (72.22)	1.79
Minority coalition (8)	3 (37.50)	1 (12.50)	1 (12.50)	2 (25.00)	5 (62.50)	2.33
Single majority (11)	11 (100)	0 (0.00)	0 (0.00)	0 (0.00)	11 (100)	2.00
Minimum-winning coalition (44)	15 (34.09)	3 (6.82)	1 (2.27)	15 (34.09)	19 (43.18)	1.66
Supermajority (18)	2 (12.50)	3 (18.75)	2 (12.50)	3 (18.75)	7 (43.75)	1.77
All cases (97)	43 (44.33)	8 (8.25)	4 (4.12)	24 (24.74)	55 (56.70)	1.81

Given that the model is able to produce a prediction, the conditional probability that the government is correctly identified is 67 %, and the conditional probability that the first maximal level identifies the government is 55.6 %. Accounting for the cases with no prediction, allows for greater confidence in the model.

Two implications can be drawn from these results. First, the relative success of the fuzzy model decreases as actual government formations become more complicated. One reason the model predicts single-majority and minority governments so well could be the model's necessity to include more parties in first-level maximal set to meet the majority-rule criterion. Because the solution set is a power set of a possible coalition, this leads to many smaller parties and coalitions being consider as possible minority governments. This contrasts with the complexity of supermajority coalitions where governments include parties unnecessary for majority control of parliaments. As the number of parties increases in actual governments, there is less likelihood of preference intersection in the model. Accordingly, the success rate of super-majority governments is significantly lower than minority governments.

The second implication is the relative success of the fuzzy maximal set. The highest maximal coalition correctly predicts the actual government in 44 % of the cases. However, when the model correctly makes a prediction, the highest maximal coalition contains the actual government in 82.8 % of cases. This relatively high success rate demonstrates that the fuzzy maximal set can reduce the number of possible coalitions with some degree of accuracy.

Finally, it is important to consider the prediction rates by country rather than government type. Table 6.5 shows the prediction rate of the model broken down by country. Only countries possessing more than nine processed elections are included because the country with next greatest number of election was Spain with only five cases. The model was most accurate for Germany and Luxembourg, correctly predicting 83.3 and 55.5 % of real-world governments, respectively. The model was least accurate for the Netherlands and Belgium correctly predicting 11.8 and 33.3 %

Table 6.5 Summary of results for 2-D analysis by country

Country (# of cases)	First-level maximal (%)	Other-level maximal (%)	No prediction (%)	% correct	# of levels
Austria (10)	20.0	20.0	30.0	40.0	2.9
Belgium (9)	33.3	0.0	33.3	33.3	2.7
Germany (12)	58.3	25.0	8.3	83.3	3.1
Luxembourg (9)	55.5	0.0	22.2	55.5	2.1
Netherlands (13)	5.9	5.9	53.0	11.8	2.8
Mean of countries	34.6	10.2	29.4	44.8	3.0

or real-world governments respectively. Considering the differences between these two sets of countries, there is an obvious relationship between the two. Germany and Luxembourg have, on average, less than five effective political parties (Golder 2005). In contrast, both Belgium and the Netherlands average more than five effective political parties. Across all 94 observations, the correlation between the accuracy of the model and the number of effective political parties is -0.27, suggesting there is a moderately strong relationship between the two variables.[17] This strengthens the earlier discussion that the model becomes less accurate as the number of actors increases.

6.5 Conclusions and Implications

This expanded and improved fuzzy model of government formation constrains the effects of intransitivity in two-dimensional space using indifference in players' preferences instead of relying on institutional restrictions or complex bargaining arguments. The results presented here give reason for confidence in the model and the fuzzy maximal set's ability to correctly identify the real-world government as the most viable coalition. This success is relevant for scholars using public choice models, most notably those who use preferences to predict outcomes outside the bounds of institutions and other formal rules. Fuzzy preferences may be a useful tool in restricting policy alternatives and predicting stable solutions.

Furthermore, the findings have empirical and theoretical implications for future research. Empirically, the model can be used to operationalize variables for statistical analysis of government formation and duration. Formal prediction sets have traditionally failed to be statistically significant in regression and event history analyses. However, the preliminary success of the fuzzy maximal set warrants further explorations of its relationship to coalition creation and maintenance in parliamentary democracies. In addition, the fuzzy approach presented here provides new validity to the CMP data set in light of recent criticism (e.g., Benoit et al. 2009) that the CMP

---

[17] Since neither variables are normally distributed, it may be more appropriate to use Spearman's rho. In this case, $r = -0.31$.

data set contains too much error in party positions. Fuzzy sets allow researchers to model preferences with data containing significant degrees of variance.

Theoretically, there are two noticeable points of contention in arguments set forth here. First, we proposed a model that is free from institutional restraints; in doing so, it has imposed restraints of its own. All fuzzy preferences profiles are treated in a similar, Likert-esque manner even though players may possess differing levels of granularity. In addition, players may not even perceive the fine distinction between α-levels, requiring some type of fuzzy indifference curve or boundary. Thus, there is a pressing need to devise a procedure to empirically infer the granularity of a player's fuzzy preference profile. Secondly, we have assumed parties act in consensus when forming coalitions. While the use of the minimum aggregator seems like a logical extension and the results suggest that it possesses some degree of validity, it may not best capture the behavior in a majority-rule environment such as parliaments. This issue of majority-rule behavior is given consideration in the next chapter, where an improved iteration of this model allows includes a weighted component, reflecting the percentage of seats a party holds in the legislature.

References

Austen-Smith, D., Banks, J.: Positive Political Theory I. The University of Michigan Press, Ann Arbor (1999)

Baron, D.P.: A spatial bargaining theory of government formation in paliamentary systems. Am. Polit. Sci. Rev. **85**(1), 137–164 (1991)

Benoit, K., Laver, M., Mikhalov, S.: Treating words as data with error: estimating uncertainty in the comparative manifesto measures. Am. J. Polit. Sci. **53**(2), 49–513 (2009)

Brown, E., Feste, K.: Qualitative dimensions of coalition payoffs: evidence for european party governments. Am. Behav. Sci. **18**(4), 530–556 (1975)

Budge, I., Robertson, D., Hearl, D.: Ideology, Strategy, and Party Change: Spatial Analyses of Post-War Election Programmes in 19 Democracies. Cambrige University Press, New York (1987)

Casey, P., Wierman, M.J., Gibilisco, M.B., Mordeson, J.N., Clark, T.D.: Assessing policy stability in iraq: a fuzzy approach to modeling preferences. Public Choice **151**(3–4), 402–423 (2012)

Clark, T.D., Larson, J., Mordeson, J.N., Potter, J., Wierman, M.J.: Applying Fuzzy Mathematics to Formal Model in Comparative Politics. Springe, Berlin (2008)

Dasgupta, M., Deb, R.: Fuzzy choice functions. Soc. Choice Welfare **8**, 171–182 (1991)

Diermeier, D., Merlo, A.: Government turnover in parliamentary democracies. J. Econ. Theor. **94**, 46–79 (2000)

Elff, M.: A spatial model of electoral platforms. Presented at the annual meeting of the midwest political science association 67th annual national conference, The Palmer House Hilton, Chicago, IL (2009)

Georgescu, I.: The similarity of fuzzy choice functions. Fuzzy Sets Syst. **158**, 1314–1326 (2007)

Golder, M.: Democratic electoral systems around the world, 1946–2000. Elect. Stud. **24**(1), 102–121 (2005)

Grofman, B.: A dynamic model of protocoalition formation in ideological n-space. Behav. Sci. **27**(1), 77–90 (1982)

Klement, E.P., Mesiar, R., Pap, E.: Triangular Norms. Kluwer Academic Publishers, Norwell (2000)

Laver, M.: How should we estimate the policy positions of political actors. In: Laver, M. (ed.) Estimating the Policy Position of Political Actors. Routledge, London (2001)

Laver, M., Schofield, N.: Multiparty Government: The Politics of Coalition in Europe. Oxford University Press, New York (1990)

Laver, M., Budge, I. (eds.): Party Policy and Government Coalitions. MacMillan Press, London (1992)

Laver, M., Hunt, B.: Policy and Party Competition. Routledge, London (1992)

Laver, M., Budge, I.: The policy basis of government coalitions: a comparative investigation. Br. J. Polit. Sci. 23(4), 499–519 (1993)

Laver, M., Shepsle, K.: Making and Breaking Government: Cabinets and Legislatures in Parliamentary Democracies. Cambrige University Press, New York (1996)

McKelvey, R.D.: Intransitives in multidimensional voting models and some implications for agenda control. J. Econ. Theor. 12(3), 472–482 (1976)

Mordeson, J.N., Clark, T.D., Miller, N.R., Casey, P.C., Gibilisco, M.B.: The uncovered set and indifference in spatial models: a fuzzy set approach. Fuzzy Sets Syst. 168, 89–101 (2011)

Riker, W.: The Theory of Political Coalitions. Yale University Press, New Haven (1962)

Schofield, N.: Political competition and multiparty coalition governments. Eur. J. Polit. Res. 23(1), 1–33 (1993)

Schucany, W.R., Polansky, A.M.: Kernal smoothing to improve bootstrap confidence intervals. Roy. Stat. Soc. 59(4), 831–838 (1997)

Van der Brug, W.: Analyzing party dynamics by taking partially overlapping snapshots. In: Laver, M. (ed.) Estimating the Policy Positions of Political Actors. Routledge, London (2001)

Wand, M.P., Jones, M.C.: Kernal Smoothing. Chapman and Hall, London (1995)

Chapter 7
The Beginnings of a Weighted Model and New Frontiers

Abstract We present some preliminary ideas on the inclusion of a weighted maximal set public choice model. We then compare the results of the tests of the models developed and tested in this book. We conclude with a consideration of fuzzy social choice functions as a means for predicting outcomes in public choice models.

7.1 A Weighted Fuzzy Government Formation Model

As discussed in the concluding remarks of the previous chapter, the results in the two-dimensional government formation model still leave room for improvement. In this chapter, we present some preliminary ideas on the use of a weighted version of the maximal set for single-dimensional *rile* CMP data. This new weighted component allows for the percentage of legislative seats held by each party to influence the model.

For this model, let $N = \{1, \ldots, n\}$ be the set of political parties and $X \subset \mathbb{R}^1$ be a set of alternatives. It is assumed each political party possesses a fuzzy choice function σ_i mapped into the interval $[0, 1]$, or $\sigma_i : X \to [0, 1]$. The model also assumes that σ_i is normal, requiring the existence of an $x \in X$ such that $\sigma_i(x) = 1$. In addition, let σ denote an n-tuple of fuzzy preference functions where $\sigma = (\sigma_1, \ldots, \sigma_n)$.

Expanding σ_i from political parties' choice functions to determining coalition formation, Gibilisco et al. use the average rule to calculate the degree to which a coalition is maximal.

Definition 7.1 For any coalition $C \subseteq N$ such that $C = \{1, \ldots, c\}$, the degree to which a coalition is maximal for a given preference profile σ is denoted by $M(\sigma)(C)$, and w_i is the legislative weight of $i \in N$, determined by the percentage of seats party i holds in the legislature.

$$M(\sigma)(C) = \max_{x \in X} \left[\sum_{i=1}^{c} w_i \cdot \sigma_i(x) \right]$$

P. C. Casey et al., *Fuzzy Social Choice Models*, Studies in Fuzziness and Soft Computing 318, DOI: 10.1007/978-3-319-08248-6_7, © Springer International Publishing Switzerland 2014

The same bootstrapping procedure used in the fuzzy maximal set single-dimensional model determines party preference profiles. The above weighted equation is then calculates the level of maximality a potential coalition possesses, mapped to the interval of [0, 1]. The results of the weighted government formation model are reported in Table 7.1. The prediction set is expanded to include all possible coalitions and the level in which the actual government formed was recorded. The average number of maximal levels recorded for this model was 23.1, but only the top four are considered when determining a correct prediction. A prediction is viewed as correct if the actual coalition formed was found in the top four maximal levels.

As shown in Table 7.1, on average, the weighted government formation model correctly predicts 60.2 % of coalition formation. This is the best average success rate of any model considered in this book. The weighted government formation model best predicts single-party majority coalitions, but at a lesser success rate of 73.1 %. However, the model shows a modest improvement in the prediction rates of minimal-winning coalitions at 68 % and super-majority coalitions at 54 %, an area where earlier models report lesser prediction rates. The weighted government formation model least accurately predicts minority coalitions with a success rate of 39.2 %.

Little work has been done on this weighted model beyond these initial results, but the high rate of prediction suggests it opens up a promising path for future research. The next natural step would be to extend the notion of a weighted maximal set to the fuzzy two-dimensional model.

7.2 A Comparison of Models

We have compared six government formation models: the conventional Median Voter model, the conventional proximity model, the fuzzy single-dimension Pareto set model, the fuzzy single-dimension maximal set model, the fuzzy weighted single-dimension maximal set model, and the fuzzy two-dimension maximal set model. Table 7.2 displays the results of the predictions made by each of these model. The level of maximality at which the model predicts the coalitions form are presented first in successive order (The Euclidean models' predictions are included in the first level maximality results as they return a single, most likely prediction for coalition formation). This is followed by statistics on the zero prediction rate and the average level of maximality for the non weighted fuzzy models. Table 7.2 also report the total number of correctly predicted coalitions and the total success rate.

No single model offers the best prediction rate across all models for each coalition type. The model with the best average prediction rate is the fuzzy weighted single-dimension maximal set model. While this removes the issue of the model making no prediction, the maximal level where the correct prediction is most often located needs further analysis.

At the first level of maximality, the fuzzy two-dimension maximal set model reports the highest prediction rate of 44.33 %. When comparing the second and

Table 7.1 Results of 1-D weighted fuzzy government formation model

Type (# of cases)	First level	Second level	Third level	Fourth level	Other level	% Correct
Single-party minority (45)	5 (11.11 %)	11 (24.44 %)	5 (11.11 %)	4 (8.89 %)	20 (44.44 %)	25 (55.55 %)
Minority coalition (28)	6 (21.42 %)	2 (7.14 %)	1 (3.57 %)	2 (7.10 %)	17 (6.07 %)	11 (39.20 %)
Single majority (26)	10 (38.46 %)	7 (26.92 %)	1 (3.85 %)	1 (3.85 %)	7 (26.92 %)	19 (73.10 %)
Minimum-winning coalition (100)	23 (23.00 %)	32 (32.00 %)	7 (7.00 %)	6 (6.00 %)	32 (32.00 %)	68 (68.00 %)
Supermajority (50)	13 (26.00 %)	4 (8.00 %)	8 (16.00 %)	2 (4.00 %)	23 (23.00 %)	27 (54.00 %)
All cases (249)	57 (22.89 %)	56 (22.49 %)	22 (8.84 %)	15 (6.02 %)	99 (39.76 %)	150 (60.20 %)

Table 7.2 All results

	Single party minority	Minority coalition	Single party majority	Minimal winning coalition	Supermajority	All cases
First level						
Conventional median voter	23 (52.27%)	0 (0.00%)	23 (100%)	0 (0.00%)	0 (0.00%)	46 (19.5%)
Conventional proximity	35 (79.55%)	8 (29.63%)	23 (100%)	26 (29.21%)	6 (11.54%)	98 (41.70%)
Fuzzy 1-D pareto set	24 (53.33%)	9 (34.62%)	21 (100%)	25 (28.74%)	6 (11.54%)	88 (36.81%)
Fuzzy 1-D maximal set	11 (64.71%)	7 (25.93%)	22 (100%)	21 (23.60%)	6 (11.54%)	75 (36.62%)
Fuzzy 2-D maximal set	12 (66.67%)	3 (37.50%)	11 (100%)	15 (34.09%)	2 (12.50%)	43 (44.33%)
Fuzzy weighted 1-D maximal set	5 (11.11%)	6 (21.42%)	10 (38.46%)	23 (23.00%)	13 (26.00%)	57 (22.89%)
Second level						
Conventional median voter	N/A	N/A	N/A	N/A	N/A	N/A
Conventional proximity	N/A	N/A	N/A	N/A	N/A	N/A
Fuzzy 1-D pareto set	N/A	N/A	N/A	N/A	N/A	N/A
Fuzzy 1-D maximal set	N/A	N/A	N/A	N/A	N/A	N/A
Fuzzy 2-D maximal set	1 (5.56%)	1 (12.50%)	0 (0.00%)	1 (2.27%)	2 (12.50%)	4 (4.12%)
Fuzzy weighted 1-D maximal set	11 (24.44%)	2 (7.14%)	7 (26.92%)	32 (32.00%)	4 (8.00%)	56 (22.49%)

(continued)

Table 7.2 (continued)

	Single party minority	Minority coalition	Single party majority	Minimal winning coalition	Supermajority	All cases
Third level						
Conventional median voter	N/A	N/A	N/A	N/A	N/A	N/A
Conventional proximity	N/A	N/A	N/A	N/A	N/A	N/A
Fuzzy 1-D pareto set	N/A	N/A	N/A	N/A	N/A	N/A
Fuzzy 1-D maximal set	N/A	N/A	N/A	N/A	N/A	N/A
Fuzzy 2-D maximal set	N/A	N/A	N/A	N/A	N/A	N/A
Fuzzy weighted 1-D maximal set	5 (11.11 %)	1 (3.57 %)	1 (3.85 %)	7 (7.00 %)	8 (16.00 %)	22 (8.84 %)
Fourth level						
Conventional median voter	N/A	N/A	N/A	N/A	N/A	N/A
Conventional proximity	N/A	N/A	N/A	N/A	N/A	N/A
Fuzzy 1-D pareto set	N/A	N/A	N/A	N/A	N/A	N/A
Fuzzy 1-D maximal set	N/A	N/A	N/A	N/A	N/A	N/A
Fuzzy 2-D maximal set	N/A	N/A	N/A	N/A	N/A	N/A
Fuzzy weighted 1-D maximal set	4 (8.89 %)	2 (7.10 %)	1 (3.85 %)	6 (6.00 %)	2 (4.00 %)	15 (6.02 %)

(continued)

Table 7.2 (continued)

	Single party minority	Minority coalition	Single party majority	Minimal winning coalition	Supermajority	All cases
Other level						
Conventional median voter	21 (47.73%)	27 (100%)	0 (0.00%)	89 (100%)	52 (100%)	189 (80.43%)
Conventional proximity	9 (20.45%)	19 (70.37%)	0 (0.00%)	63 (70.79%)	46 (88.46%)	137 (58.30%)
Fuzzy 1-D pareto set	0 (0.00%)	5 (19.23%)	0 (0.00%)	18 (20.69%)	11 (21.15%)	34 (14.72%)
Fuzzy 1-D maximal set	0 (0.00%)	5 (18.52%)	0 (0.00%)	17 (19.10%)	11 (21.15%)	30 (14.49%)
Fuzzy 2-D maximal set	0 (0%)	1 (12.50%)	0 (0.00%)	1 (2.27%)	2 (12.50%)	4 (4.12%)
Fuzzy weighted 1-D maximal set	20 (44.44%)	17 (6.07%)	7 (26.92%)	32 (32.00%)	23 (23.00%)	99 (39.76%)
No prediction						
Conventional median voter	N/A	N/A	N/A	N/A	N/A	N/A
Conventional proximity	N/A	N/A	N/A	N/A	N/A	N/A
Fuzzy 1-D pareto set	3 (6.67%)	1 (3.85%)	0 (0.00%)	13 (14.94%)	5 (9.62%)	22 (9.52%)
Fuzzy 1-D maximal set	4 (23.53%)	5 (18.52%)	0 (0.00%)	18 (20.22%)	9 (17.31%)	36 (17.39%)
Fuzzy 2-D maximal set	4 (22.22%)	2 (25.00%)	0 (0.00%)	15 (34.09%)	3 (18.75%)	24 (24.74%)
Fuzzy weighted 1-D maximal set	N/A	N/A	N/A	N/A	N/A	N/A

(continued)

Table 7.2 (continued)

	Single party minority	Minority coalition	Single party majority	Minimal winning coalition	Supermajority	All cases
Average levels						
Conventional median voter	N/A	N/A	N/A	N/A	N/A	N/A
Conventional proximity	N/A	N/A	N/A	N/A	N/A	N/A
Fuzzy 1-D pareto set	2.4	2.6	2.7	2.3	2.5	2.4
Fuzzy 1-D maximal set	2.9	3.0	3.2	2.9	3.0	3.0
Fuzzy 2-D maximal set	1.8	2.3	2.0	1.7	1.8	1.8
Fuzzy weighted 1-D maximal set	UNK	UNK	UNK	UNK	UNK	23.1
Total correct						
Conventional median voter	23 (52.27 %)	0 (0.00 %)	23 (100 %)	0 (0.00 %)	0 (0.00 %)	46 (19.57 %)
Conventional proximity	35 (79.55 %)	8 (29.63 %)	23 (100 %)	26 (29.21 %)	6 (11.54 %)	98 (41.70 %)
Fuzzy 1-D pareto set	24 (53.33 %)	14 (53.80 %)	21 (100 %)	43 (49.2 %)	17 (32.69 %)	119 (51.50 %)
Fuzzy 1-D maximal set	11 (64.71 %)	12 (44.44 %)	22 (100 %)	38 (42.70 %)	17 (32.69 %)	100 (48.31 %)
Fuzzy 2-D maximal set	13 (72.22 %)	5 (62.50 %)	11 (100 %)	19 (43.18 %)	7 (43.75 %)	55 (56.70 %)
Fuzzy weighted 1-D maximal set	25 (55.55 %)	11 (39.20 %)	19 (73.10 %)	68 (68.00 %)	27 (54.00 %)	150 (60.20 %)

higher level prediction rates between it and the fuzzy weighted single-dimension maximal set model, the precent correct for all cases should be added. With the second level included, the fuzzy two-dimension maximal set mode has a higher prediction rate. Due to its ability to consider a greater number of levels of set inclusion in the maximal set, the fuzzy weighted single-dimension maximal set model surpasses the the fuzzy two-dimension maximal set model's total prediction rate when all levels of maximality are taken in account.

The non-weighted fuzzy models also report a rate of no prediction in the cases where no overlap existed between political parties. Of the three models, the fuzzy two-dimension maximal set model reports the highest no prediction rate at 24.74 %. The fuzzy single-dimension Pareto set model reports the lowest no prediction rate of these three models at 9.52 %.

7.3 Future Directions in Fuzzy Public Choice Models

The fuzzy public choice models presented in this book offer an important alternative to the conventional approach. In particular, the fuzzy models eschew some of the more restrictive assumptions of the conventional public choice models. In general, all of the fuzzy models out-performed the conventional models. The weighted fuzzy model of coalition formation offers the best prediction rate on average. This model extends the prediction set to include the degree of maximality for every possible coalition. However, no single model offers the best prediction rate across all models for each coalition type.

All of the approaches that we have tested in this book rely on the use of an aggregation rule for ordering the fuzzy preference orders of players into a collective fuzzy preference. The social choice literature focuses instead on the effect of specific rules on what alternative is chosen. Thus, it removes the assumption that the choice is constrained to the maximal set or the Pareto set and thereby ignores the problems associated with intransitivity in social preference. Thus, fuzzy social choice offers an important direction for future empirical research.

Rational actors may not able to perceive every alternative in a set of alternatives X. They may also choose to exclude some alternatives from consideration. Let S be a subset of X that some set of political actors are not able to perceive. Then the complement $X \backslash S$ is the set comprising those alternatives considered in the decision process. If C is a choice function on X, then $C(X \backslash S)$ are the alternatives chosen and $X \backslash C(X \backslash S)$ are those that are not. This leads us to the concept of an upper choice function. In this concluding section of the book we give consideration to upper choice functions.

The study of decision-making begins with a choice function C that maps $\mathscr{P}^*(X)$ into itself such that $C(S) \subseteq S$ for all $S \in \mathscr{P}^*(X)$, where X is the universe of alternatives. In many real world applications, not every alternative is available. Restricting the universe of alternatives yields some interesting properties of choice functions which we examine here.

The notion of an upper choice function $\overline{C}$ in terms of a choice function C on a set of alternatives X is taken essentially from rough set theory Pawlak (1991). Our work is based on Mordeson et al. (2008). There is a significant difference in the study of C and $\overline{C}$ here from upper and lower approximation operators in rough set theory in that if S_1 and S_2 are subsets of X such that $S_1 \subseteq S_2$, it is not necessarily the case that $C(S_1) \subseteq C(S_2)$. Also a choice function is not allowed to be applied to the empty set and it is not necessarily the case that $C(S) = C(C(S))$ $\forall S \subseteq X$. We note in the argument that follows that C and $\overline{C}$ cannot be defined nontrivially in terms of an equivalence relation on X as is the case in rough set theory. For these reasons, the results we obtain do not follow immediately from rough set theory. Our approach is axiomatic in nature as is the approach in other situations such as successor and source functions in (fuzzy) finite state machines, (fuzzy) directed graphs (Kuroki and Mordeson 1997a, b; Mordeson 1999a, b; Malik and Mordeson 2002). With certain assumptions, the proofs of the results can be patterned after those in Mordeson (1999a). Due to the importance of choice functions, it is never-the-less important to establish these results. In Fedrizzi et al. (1996), a connection between choice functions and a rough set approach is discussed.

7.3.1 Basic Concepts

Let S be a subset of X. It is the case that rational political players are not able to perceive every alternative. Therefore, S can be interpreted as the subset of alternatives that some set of political actors are not able to perceive. In this case, the set $X \backslash S$ would make up those alternatives considered in the decision process, and $C(X \backslash S)$ would be those alternatives chosen from among those considered.

Alternatively, S can be interpreted as the set of alternatives that are perceived, but purposefully excluded from consideration. This is not an infrequent phenomenon in politics. For instance consider the possibility that trade unions must be included in drafting legislation on labor law. We could easily imagine that these unions would veto a large number of options from being considered. These options would be in S. Another example is provided by party systems. The options offered voters is necessarily reduced by some set of alternatives S, which represents the alternatives not offered by the parties. The same is the case for a whole host of institutions from legislative committees to the nation-states in the United Nations Security Council. In every case, some set of actors exercises a de facto veto, or de jure veto in case of five members of the Security Council, over alternatives considered. In such cases, $X \backslash S$ is the set of alternatives from among which the players choose, $C(X \backslash S)$ comprises the alternatives that are chosen, and $X \backslash C(X \backslash S)$ is the set of options that are neither vetoed nor chosen in the final selection process.

It is interesting that the exclusion from consideration of some set of alternatives S can result in a different outcome than might otherwise have occurred had S not been excluded. For instance, suppose that an institution that can exercise a de facto or dejure veto prefers some alternative x. Further suppose that $S = \emptyset, x, y \in X, xPy$

and that $C(X\backslash S) = C(X) = \{x\}$. Since $S = \emptyset$, the institution did not exercise the veto. Now suppose that $S \neq \emptyset$ (in which case the institution exercised the right of veto), $x, y \in X$, $x P y$ by the veto player and that $x, y \notin S$. It is possible that $C(X\backslash S) = \{y\}$. In essence, the institution that vetoed the set of alternatives in S preferred x to some alternative y, but the unintended consequence of the veto was that the resulting social choice is y, the less preferred option. Had the set S not been excluded from consideration, x would have emerged as the social choice. The same outcome adheres if the alternative was not vetoed, but was simply not perceived by the players.

We illustrate this as follows. Suppose that a committee comprising representatives from fifteen countries (such as the United Nations Security Council) is faced with a choice between four options, $X = \{x, y, z, w\}$. Suppose a group of six countries possess the strict preference order $x P_1 z P_1 y P_1 w$, a group of four possess the strict preference order $y P_2 z P_2 x P_2 w$, a group of three possess the strict preference order $z P_3 y P_3 w P_3 x$, and a group of two possess the strict preference order $w P_4 z P_4 y P_4 x$. If a majority rule is used, x receives six votes in the first round, y receives four votes, z receives three votes and w receives two votes. In the second round which pits the top two vote getters from the first round, y defeats x by a vote of nine to six. Now suppose that one of the countries in the second group has a veto and that the country uses the veto to exclude alternative w, the group's last choice from consideration. In such case, the two top vote-getters in the first round are x with six votes and z with five votes. The second group's ideal alternative, y, which would have won had w not been excluded, now loses in the first round.. The winner in the second round run-off is option z with nine votes against the six votes of option x. Hence, the elements in $X\backslash C(X\backslash S)$ are a function of the elements that are contained in S. The public choice literature demonstrates, that this counterintuitive result can occur under a wide range of rules for aggregating collective preferences from individual preferences. A discussion of this literature can be found in Austen-Smith and Banks (1999) and Austen-Smith and Banks (2005).

7.3.2 Structure Results

Let X be a nonempty set. A function C from $\mathscr{P}^*(X)$ into $\mathscr{P}^*(X)$ such that $C(S) \subseteq S$ $\forall S \in \mathscr{P}^*(X)$ is called a **choice function** on X. In this section, we develop structure results for choice functions using an algebraic approach.

Definition 7.2 Let C be a choice function on X. Then C is said to satisfy the **generating properties** if the following conditions hold:

(1) $\forall S, T \in \mathscr{P}^*(X), S \subseteq T$ implies either $C(S) \cap C(T) = \emptyset$ or $C(S) \subseteq C(T)$ (condition β Austen-Smith and Banks (1999)),
(2) $\forall S \in \mathscr{P}^*(X), C(C(S)) = C(S)$ (closure property),
(3) $\forall S \in \mathscr{P}^*(X), \forall x, y \in X, x \notin C(X\backslash(S \cup \{y\}))$ and $x \in C(X\backslash S)$ imply $y \notin C(X\backslash(S \cup \{x\}))$ (exchange property).

In words, condition β says that if any alternative in a set S that is chosen remains chosen if S is extended to a set T, then all the alternatives that were chosen from S remain chosen. The closure property says that if a player chooses alternatives x and y from a choice between x, y, and z, then when reducing the available alternatives to only x and y, the player will still choose x and y. The exchange property says that if an alternative x is chosen from the complement of a set S in X, but is not chosen if alternative y is also removed, then y is not chosen if alternative x is removed from the complement of S in X.

Definition 7.3 Let C be a choice function on X and $S \in \mathcal{P}^*(X)$. Then S is called a C-**subspace** of X if $S = C(S)$.

Note it is possible in Definition 7.3 that $C(X) \neq X$ in which case X is not a C-subspace of itself.

Proposition 7.4 *Let C be a choice function on X. Suppose C satisfies condition β and the closure property. Let $S \in \mathcal{P}^*(X)$ and $x \in X$. If $x \notin C((X \backslash S) \cup \{x\})$, then either $C((X \backslash S) \cup \{x\}) \cap C(X \backslash S) = \emptyset$ or $C((X \backslash S) \cup \{x\}) = C(X \backslash S)$.*

Proof We have

$$X \backslash ((X \backslash S) \cup \{x\}) \subseteq X \backslash C((X \backslash S) \cup \{x\})$$

since C is a choice function and

$$\{x\} \subseteq X \backslash C((X \backslash S) \cup \{x\})$$

by hypothesis. Thus

$$S \subseteq \{x\} \cup (X \backslash ((X \backslash S) \cup \{x\})) \subseteq X \backslash C((X \backslash S) \cup \{x\}).$$

Hence

$$X \backslash S \supseteq C((X \backslash S) \cup \{x\}).$$

Thus either

$$C((X \backslash S) \cup \{x\}) \cap C(X \backslash S) = \emptyset$$

or

$$C(X \backslash S) \supseteq C(C((X \backslash S) \cup \{x\})) = C((X \backslash S) \cup \{x\})$$

by the closure property. Now either

$$C((X \backslash S) \cup \{x\}) \cap C(X \backslash S) = \emptyset$$

or

$$C(X \backslash S) \subseteq C((X \backslash S) \cup \{x\})$$

by condition β. Hence either

$$C((X \backslash S) \cup \{x\}) \cap C(X \backslash S) = \emptyset$$

or

$$C(X \backslash S) = C((X \backslash S) \cup \{x\}).$$

Definition 7.5 Let C be a choice function on X and $S \in \mathscr{P}^*(X)$. Then S is called C-**free** if $\forall x \in S, x \in C((X \backslash S) \cup \{x\})$. Let U be a C-subspace of X. Then S is said to be C-**free with respect to** U if S is C-free and $S \subseteq X \backslash U$. Even though $\emptyset \notin \mathscr{P}^*(X)$, we sometimes allow $\emptyset$ since $\emptyset$ would be C-free vacuously.

Definition 7.6 Let C be a choice function on X and $S \in \mathscr{P}^*(X)$. Let U be a C-subspace of X. Then S is said to C-**generate** U if $U = C(X \backslash S)$.

Let U be a C-subspace of X. Suppose that $S \in \mathscr{P}^*(X)$ C-generates U. Then $U = C(X \backslash S) \subseteq X \backslash S$. Thus $U \cap S = \emptyset$. In particular, if X is a C-subspace of itself, then $S = \emptyset$, a contradiction.

A C-generating set has the property that as the set of alternatives that when removed, allows some other set of alternatives to be chosen. For example, if a political actor chooses alternatives x and y when u and v are unavailable, u and v can be said to C-generate x and y.

Definition 7.7 Let C be a choice function on X and $S \in \mathscr{P}^*(X)$. Then S is called **maximal** C-**free** if S is C-free and $\forall z \in X \backslash S, S \cup \{z\}$ is not C-free. Let U be a C-subspace of X. Then S is called **maximal** C-**free** with respect to U if S is maximal C-free and $S \subseteq X \backslash U$.

It follows that $S \in \mathscr{P}^*(X)$ is maximal C-free if and only if S is C-free and

$$\forall z \in X \backslash S \; \exists y \in S \cup \{z\} \; y \notin C\left((X \backslash (S \cup \{z\})) \cup \{y\}\right).$$

Definition 7.8 Let U be a C-subspace of X. Let $S \in \mathscr{P}^*(X)$. Then S is called a **minimal** C-**generating set** for U if S C-generates U and $\forall S' \subset S, S'$ does not C-generate U, i.e., $C(X \backslash S) \subset C(X \backslash S')$.

Definition 7.9 Let U be a C-subspace of X. Let $S \in \mathscr{P}^+(X)$. Then S is called a C-**basis** of U if S C-generates U and S is C-free with respect to U.

Example 7.10 Let $X = \{x, y, z\}$. Define $C : \mathscr{P}^*(X) \to \mathscr{P}^*(X)$ by $\forall S \in \mathscr{P}^*(X)$, $C(S) = S \backslash \{z\}$ if $S \neq \{z\}$ and $C(\{z\}) = \{z\}$. Then C is a choice function on X. Now $C(S) = S$ if $S = \{z\}$ or $z \notin S$. Hence condition β and the closure property clearly hold.

Let $S \in \mathscr{P}^*(X), u, v \in X, u \notin C(X \backslash (S \cup \{v\}))$ and $u \in C(X \backslash S)$, where $X \backslash (S \cup \{v\}) \neq \emptyset$. Since $u \in C(X \backslash S), v \notin S$. Thus $|(X \backslash S) \cup \{v\}| = 2$. Hence $|X \backslash (S \cup \{v\})| = 1$. Thus $C(X \backslash (S \cup \{v\})) = X \backslash (S \cup \{v\})$. Hence $u \notin X \backslash (S \cup \{v\})$ and $u \in C(X \backslash S) \subseteq X \backslash S$. Thus $u = v$. Hence $v \notin X \backslash (S \cup \{u\}) = C(X \backslash (S \cup \{u\}))$. Thus the exchange property holds.

Now $C(X) = \{x, y\} = C(X\backslash\{z\})$, but $\{z\}$ is not C-free since $z \notin \{x, y\} = C((X\backslash\{z\}) \cup \{z\})$. Let $S = \{y, z\}$. Then $\{x\} = C(X\backslash\{y, z\})$, but $z \notin \{x\} = C((X\backslash S) \cup \{z\})$. That is, S is a C-generating set for $\{x\}$, but S is not C-free with respect to $\{x\}$. It follows that $\{y\}$ is a C-basis for $\{x\}$ since $y \in \{x, y\} = C((X\backslash\{y\}) \cup \{y\})$, $\{y\} \subseteq X\backslash\{x\}$, and $\{x\} = C(X\backslash\{y\})$. We note that $\{z\} \subseteq \{y, z\}$, $C(\{z\}) \nsubseteq C(\{y, z\})$, and in fact $C(\{z\}) \cap C(\{y, z\}) = \emptyset$. We also note that R rationalizes C, Fedrizzi et al. (1996), where $R = \{(x, x), (y, y), (z, z), (x, y), (y, x), (x, z), (y, z)\}$.

Proposition 7.11 *Let C be a choice function on X. Suppose C satisfies the exchange property. If $S \in \mathscr{P}(X)$ is C-free and $z \in C(X\backslash S)$, then $S \cup \{z\}$ is C-free.*

Proof Suppose there exists $x \in S$ such that $x \notin C((X\backslash(S \cup \{z\})) \cup \{x\})$. Since S is C-free, $x \in C((X\backslash S) \cup \{x\})$. It follows that $C((X\backslash(S \cup \{z\})) \cup \{x\}) = C((X\backslash(S\backslash\{x\}) \cup \{z\}))$ and $C((X\backslash S) \cup \{x\}) = C(X\backslash(S\backslash\{x\}))$. Thus by (3) of Definition 7.2, $z \notin C((X\backslash(S\backslash\{x\}) \cup \{x\}) = C(X\backslash S)$, a contradiction. Hence $\forall x \in S, x \in C((X\backslash(S \cup \{z\})) \cup \{x\}))$. We also have that $z \in C((X\backslash(S \cup \{z\})) \cup \{z\})) = C(X\backslash S)$ by hypothesis. Thus $S \cup \{z\}$ is C-free.

Example 7.12 Let X and C be defined as Example 7.10. Then the following table lists the C-subspaces of X and their basis.

C-subspaces	$\{x\}$	$\{y\}$	$\{z\}$	$\{x, y\}$
C-basis	$\{y\}$	$\{x\}$	$\{x, y\}$	no basis or $\emptyset$

Consider the C-subspace $\{x, y\}$. As shown in Example 7.10, $\{z\}$ is not C-free. Thus $\{z\}$ is not a basis for $\{x, y\}$. It follows that $\{x, y\}$ has no basis unless $\emptyset$ is allowed. Note that $\{x, y\} = C(X\backslash\emptyset)$. That is, $\emptyset$ C-generates $\{x, y\}$. Consider the C-subspace $\{z\}$. Then

$$x \in C((X\backslash\{x, y\}) \cup \{x\}), y \in C((X\backslash\{x, y\}) \cup \{y\}), \{x, y\} \subseteq X\backslash\{z\},$$

and $\{z\} = C(X\backslash\{x, y\})$. Hence $\{x, y\}$ is a basis for $\{z\}$. Consider the C-subspace $\{y\}$. Then $x \in X((X\backslash\{x\}) \cup \{x\})$, $\{x\} \subseteq X\backslash\{y\}$, and $\{y\} = C(X\backslash\{x\})$. Thus $\{x\}$ is a C-basis for $\{y\}$. For the C-subspace $U = \{z\}$, $S = \{x\}$ is C-free with respect to U, and $S \cup S'$ is C-free with respect to U, where $S' = \{y\}$. We have

$$C(U) \cap C(X\backslash S) = \{z\} \cap \{y\}$$
$$= \emptyset,$$
$$C(U) \cap C(X\backslash S \cup S') = \{z\} \cap \{z\}$$
$$\neq \emptyset.$$

Thus the assumption in the following proposition does not hold, but the conclusion never-the-less holds. Now let $U = \{x\}$. Let $S = \emptyset$ and $S' = \{y\}$. Then S and $S \cup S'$ are C-free with respect to U. Now $C(U) \cap C(X\backslash S) = \{x\} \cap \{x, y\} \neq \emptyset$ and $C(U) \cap C(X\backslash S \cup S') = \{x\} \cap \{x\} \neq \emptyset$. Hence the assumption in Proposition 7.14 below holds.

Example 7.13 Let $X = \{x, y, z\}$. Define $C : \mathscr{P}^*(X) \to \mathscr{P}^*(X)$ as follows:

$$C(X) = \{x, y\}, C(\{x, y\}) = \{x\}, C(\{x\}) = \{x\},$$
$$C(\{x, z\}) = \{z\}, C(\{z\}) = \{z\}, C(\{y, z\}) = \{z\}, C(\{y\}) = \{y\}.$$

Then $C(X\backslash\{z\}) = \{x\}$. Thus $\{z\}$ C-generates $\{x\}$. Now $z \notin \{x, y\} = C((X\backslash\{z\}) \cup \{z\})$. Thus $\{z\}$ is not C-free. Also $y \in C(X\backslash\{y\})\cup\{y\})$ and so $\{y\}$ is C-free. However, $C(X\backslash\{y\}) \neq \{x\}$. Thus $\{y\}$ does not C-generate $\{x\}$. Thus it follows that $\{x\}$ does not have a C-basis. (Note also that $\emptyset \subseteq X\backslash\{x\}$, but $\{x\} \neq C(X\backslash\emptyset)$ so $\{x\}$ doesn't have $\emptyset$ for a C-basis even if we allowed $\emptyset$.)

Proposition 7.14 *Let C be a choice function on X satisfying the generating properties. Let U be a C-subspace of X. Let S be a subset of X such that S is C-free with respect to U. Suppose that $C(U) \cap C(X\backslash S^*) \neq \emptyset$ for any subset S^* of X such that $S^* \supseteq S$ and S^* is C-free with respect to U. Suppose there exists a (finite) subset Y of X such that $C(X\backslash Y) = U$. Then there exists $S' \subseteq Y$ such that $S \cup S'$ is C-basis for U and $S \cap S' = \emptyset$.*

Proof There exists $S'' \subseteq Y$ such that $S \cup S''$ is C-free with respect to U and $S \cap S'' = \emptyset$, namely $S'' = \emptyset$. Since Y is finite, there exists a maximal subset S' of Y such that $S \cup S'$ is free with respect to U and $S \cap S' = \emptyset$. In order to show $S \cup S'$ is a C-basis for U, it remains to show that $U = C(X\backslash S \cup S')$. Since $S \cup S' \subseteq X\backslash U, U \subseteq X - S \cup S'$ and so $U = C(U) \subseteq C(X\backslash S \cup S')$ by condition β since $C(U) \cap C(X\backslash S \cup S') \neq \emptyset$ by hypothesis. Suppose $Y \cap C(X\backslash S \cup S') = \emptyset$. Then $X\backslash Y \supseteq C(X\backslash S \cup S')$. Hence $U = C(X\backslash Y) \supseteq C(C(X\backslash S \cup S') = C(X\backslash S \cup S')$ by the closure property. Hence $U = C(X\backslash S \cup S')$. Suppose $Y \cap C(X\backslash S \cup S') \neq \emptyset$. Then there exists $z \in C(X\backslash S \cup S') \cap Y$ such that $z \notin U$. Then $S \cup S' \cup \{z\}$ is C-free by Proposition 7.11. Since $z \notin U, (S \cup S' \cup \{z\}) \cap U = \emptyset$. Also $z \in C(X\backslash S \cup S') \subseteq X\backslash S \cup S' \subseteq X\backslash S$ and so $z \notin S$. Hence $S \cap (S' \cup \{z\}) = \emptyset$, a contradiction of the maximality of S'. Thus $U = C(X\backslash S \cup S')$.

Corollary 7.15 *Let C be a choice function on X satisfying the generating properties. Let U be a subspace of X. Suppose $C(U) \cap C(X\backslash S^*) \neq \emptyset$ for all subset S^* of X such that S^* is C-free with respect to U. If U is finitely generated, then U has a C-basis.*

Proof The result follows by letting $S = \emptyset$ in Proposition 7.3.

Proposition 7.16 *Let C be a choice function on X and let U be a C-subspace of X. Let S be a subset of X. Suppose conditions (a) and (b) hold.*

(a) *If $x \notin C((X\backslash S) \cup \{x\})$, then $C((X\backslash S) \cup \{x\}) \cap C(X\backslash S) \neq \emptyset$.*
(b) *If U is a C-subspace of X and $U \subseteq X\backslash S$, then $C(U) \cap C(X\backslash S) \neq \emptyset$.*

If C satisfies condition β and the closure property, then the following assertions are equivalent.

(1) *S is a minimal C-generating set of U.*
(2) *S is a maximal C-free subset of X with respect to U.*
(3) *S is a C-basis of U.*

Proof (1) $\Rightarrow$ (3) : Suppose S is not C-free. Then there exists $x \in S$ such that $x \notin C((X \backslash S) \cup \{x\})$. Hence $C((X \backslash S) \cup \{x\}) = C(X \backslash S)$ by Proposition 7.11. Thus $C(X \backslash (S \backslash \{x\})) = C(X \backslash S)$ which contradicts $C(X \backslash (S \backslash \{x\})) \supset C(X \backslash S)$ since S is C-minimal. Hence X is C-free. Since $U = C(X \backslash S) \subseteq X \backslash S, X \backslash U \supseteq S$. Thus S is C-free with respect to U.

(3) $\Rightarrow$ (2) : We have that S is C-free and $U = C(X \backslash S)$. Suppose that there exists $z \in X \backslash S, z \notin U$, such that $S \cup \{z\}$ is C-free. If $z \notin C(X \backslash S)$, then $z \notin C((X \backslash (S \cup \{z\})) \cup \{z\})$ and so $S \cup \{z\}$ is not C-free, a contradiction. Hence $z \in C(X \backslash S) = U$, a contradiction. Thus z does not exist and so S is maximal with respect to U.

(2) $\Rightarrow$ (1) : Since S is a maximal C-free subset of X with respect to U, $S \cap U = \emptyset$ and so $U \subseteq X \backslash S$. Hence $U = C(U) \subseteq C(X \backslash S)$ by condition β. Suppose that $z \notin U$ and $z \in C(X \backslash S)$. Then $z \in X \backslash S$. By the maximality of S, $z \notin C(X \backslash (S \cup \{z\})) \cup \{z\}) = C(X \backslash S)$, where the equality hold since $z \notin S$. However this is impossible. Thus z doesn't exist and so $U = C(X \backslash S)$. Suppose that there exists $x \in S$ such that $U = C(X \backslash (S \backslash \{x\}))$. Since S is C-free, $x \in C((X \backslash S) \cup \{x\}) = C(X \backslash (S \backslash \{x\})) = U$, a contradiction of the fact that $U \cap S = \emptyset$ since S is C-free with respect to U. Hence S is minimal.

The proof of the following result is similar to that of Theorem 13 in Mordeson (1999a).

Theorem 7.17 *Let C be a choice function on X satisfying the generating properties. Let U be a C-subspace of X. Suppose that C satisfies the properties of Proposition 7.16. If U has a C-basis of n elements, then every C-basis of U has n elements.*

A C-subspace is a set of alternatives whose members do not prevent the other alternatives in the subspace being chosen. If x and y are in the subspace, x does not prevent y from being chosen and conversely. A set that C-generates a subspace is the set of alternatives which, when removed, make the choice of the subspace possible. A set that is C-free contains elements that do not rely on the presence of other alternatives in the set to be chosen. If x and y are in a C-free set, then omitting x does not prevent y from being chosen and omitting y does not prevent x from being chosen.

7.3.3 Upper Choice Functions

We now define an upper choice function in terms of a choice function. We determine relationships between choice functions and upper choice functions. Let $\mathscr{P}^-(X) = \mathscr{P}(X) \backslash \{X\}$.

Definition 7.18 Let C be a choice function on X. Define $\overline{C} : \mathscr{P}^-(X) \to \mathscr{P}^-(X)$ by $\forall S \in \mathscr{P}^-(X), \overline{C}(S) = X \backslash C(X \backslash S)$. Then $\overline{C}$ is called an **upper choice function** on X.

Let C be a choice function on X and $\overline{C}$ the corresponding upper choice function on X. The choice function C is sometimes referred to as a **lower choice function.** It is immediate that $C(S) = X \backslash \overline{C}(X \backslash S) \forall S \in \mathscr{P}^*(X)$. It is also immediate that $C(S) \subseteq \overline{C}(S) \forall S \in \mathscr{P}^*(X) \cap \mathscr{P}^-(X)$,

$$x \in C(S) \Rightarrow x \in S \Leftrightarrow x \notin X \backslash S \Rightarrow x \notin C(X \backslash S)$$
$$\Leftrightarrow x \in X \backslash C(X \backslash S)$$
$$\Leftrightarrow x \in \overline{C}(S).$$

Let $S \in \mathscr{P}^*(X) \cap \mathscr{P}^-(X)$. Then $\overline{C}(S) \backslash C(S)$ is called the **boundary** of S.

Definition 7.19 Let C be a choice function on X and $\overline{C}$ the corresponding upper choice function on X. Let $S \in \mathscr{P}^-(X)$. Then S is called a $\overline{C}$-**subspace** of X if $\overline{C}(S) = S$.

Proposition 7.20 *Let C be a choice function on X and $\overline{C}$ the corresponding upper choice function on X. Suppose that C is rationalized by the relation R on X. Then $\forall S \in \mathscr{P}^*(X) \cap \mathscr{P}^-(X)$,*

$$\overline{C}(S) = C(S)$$
$$\cup \{x \in S \mid \exists y \in S \ not \ xRy\}$$
$$\cup \{x \in X \backslash S \mid \exists y \in X \backslash S \ not \ xRy\}.$$

Proof It follows that

$$\overline{C}(S) = X \backslash C(X \backslash S)$$
$$= X \backslash \{x \in X \backslash S \mid xRy \ \forall y \in X \backslash S\}$$
$$= S \cup \{x \in X \backslash S \mid not \ (xRy \forall y \in X \backslash S)\}$$
$$= S \cup \{x \in X \backslash S \mid \exists y \in X \backslash S, \ not \ xRy\}$$
$$= C(S)$$
$$\cup \{x \in S \mid \exists y \in S \ not \ xRy\}$$
$$\cup \{x \in X \backslash S \mid \exists y \in X \backslash S \ not \ xRy\}.$$

Let $S \in \mathscr{P}^*(S)$. Then S is a C-subspace of X if and only if $X \backslash S$ is a $\overline{C}$-subspace of X. Let $S \in \mathscr{P}^-(X)$. Then S is a $\overline{C}$-subspace of X if and only if $X \backslash S$ is a C-subspace of X.

Example 7.21 Consider Examples 7.10 and 7.13. Then

$$\overline{C}(\{x, z\}) = X \backslash C(X \backslash \{x, z\})$$
$$= X \backslash C(\{y\})$$
$$= X \backslash \{y\}$$
$$= \{x, z\}.$$

Thus $\{x, z\}$ is a $\overline{C}$-subspace of X. Also,

$$\{x\} = C(\{x\}),$$
$$\{z\} = \{w \in \{x, z\} \mid \text{not } wR_C u \ \forall u \in \{x, z\}\}, \text{ and}$$
$$\emptyset = \{w \in X \backslash \{x, z\} \mid \text{not } wR_C u \ \forall u \in X \backslash \{x, z\}\}$$

where R_C is the base relation associated with C, Austen-Smith and Banks (1999).

Proposition 7.22 *Let C be a choice function on X and $\overline{C}$ the corresponding upper choice function on X. Then the following assertions hold*

(1) *Suppose $S, T \in \mathscr{P}^*(X)$.*
Then $C(S) \cap C(T) = \emptyset$ if and only if $\overline{C}(X \backslash S) \cup \overline{C}(X \backslash T) = X$.
(2) *Suppose $S, T \in \mathscr{P}^-(X)$.*
Then $\overline{C}(S) \cup \overline{C}(T) = X$ if and only if $C(X \backslash S) \cap C(X \backslash T) = \emptyset$.
(3) *Suppose $S, T \in \mathscr{P}^*(X)$.*
Then $C(S) \subseteq C(T)$ if and only if $\overline{C}(X \backslash S) \supseteq \overline{C}(X \backslash T)$.
(4) *Suppose $S, T \in \mathscr{P}^-(X)$.*
Then $\overline{C}(S) \subseteq \overline{C}(T)$ if and only if $C(X \backslash S) \supseteq C(X \backslash T)$.

The following definitions are suggested by the preceding result.

Definition 7.23 Let C be a choice function on X and $\overline{C}$ the corresponding upper choice function on X. Then

(1) $\overline{C}$ is said to satisfy condition β if $\forall S, T \in \mathscr{P}^-(X)$,

$$S \subseteq T \text{ implies } \overline{C}(S) \cup \overline{C}(T) = X \text{ or } \overline{C}(S) \subseteq \overline{C}(T).$$

(2) $\overline{C}$ is said to satisfy condition α if $\forall S, T \in \mathscr{P}^-(X)$,

$$S \subseteq T \text{ implies } \overline{C}(S) \cup T \supseteq \overline{C}(T).$$

(3) $\overline{C}$ is said to satisfy condition γ if $\forall S, T \in \mathscr{P}^-(X)$,

$$\overline{C}(S) \cup \overline{C}(T) \supseteq \overline{C}(S \cap T).$$

(4) $\overline{C}$ is said to satisfy Arrow if $\forall S, T \in \mathscr{P}^-(X)$,

$$S \subseteq T \text{ implies } T \cup \overline{C}(S) = X \text{ or } T \cup \overline{C}(S) = \overline{C}(T).$$

(5) $\overline{C}$ is said to satisfy the weak axiom of revealed preferences (WARP) if $\forall S, T \in \mathscr{P}^-(X)$,

$$x \notin \overline{C}(S), \ y \in \overline{C}(S) \backslash S \text{ and } y \notin \overline{C}(T) \text{ implies } x \in T.$$

(6) $\overline{C}$ is said to satisfy Pareto Indifference (PI) if $\forall S, T \in \mathscr{P}^-(X)$,

$$\overline{C}(S \cap T) = \overline{C}(\overline{C}(S) \cap \overline{C}(T)) .$$

The next result shows the strong relationship between these conditions with respect to C and $\overline{C}$.

Proposition 7.24 *Let C be a choice function on X and $\overline{C}$ the corresponding upper choice function on X. Then the following assertions hold.*

(1) $\overline{C}$ satisfies condition β if and only if C satisfies condition β.
(2) $\overline{C}$ satisfies condition α if and only if C satisfies condition α.
(3) $\overline{C}$ satisfies condition γ if and only if C satisfies condition γ.
(4) $\overline{C}$ satisfies condition Arrow if and only if C satisfies condition Arrow.
(5) $\overline{C}$ satisfies condition WARP if and only if C satisfies condition WARP.
(6) $\overline{C}$ satisfies condition PI if and only if C satisfies condition PI.

Proof In the following proofs we always assume that $S, T \in \mathscr{P}^*(X)$,

(1) C satisfies condition $\beta \Leftrightarrow \overline{C}$ satisfies condition β. We have for $\forall S, T \in \mathscr{P}^*(X)$,

$$\begin{aligned}
C \text{ satisfies condition } \beta &\Leftrightarrow S \subseteq T \Rightarrow C(S) \cap C(T) = \emptyset \\
&\quad \text{or } C(S) \subseteq C(T) \\
&\Leftrightarrow S \subseteq T \Rightarrow (X \backslash \overline{C}(X \backslash S)) \cap (X \backslash \overline{C}(X \backslash T)) = \emptyset \\
&\quad \text{or } X \backslash \overline{C}(X \backslash S) \subseteq X \backslash \overline{C}(X \backslash T) \\
&\Leftrightarrow S \subseteq T \Rightarrow X \backslash (\overline{C}(X \backslash S) \cup \overline{C}(X \backslash T)) = \emptyset \\
&\quad \text{or } \overline{C}(X \backslash S) \supseteq \overline{C}(X \backslash T) \\
&\Leftrightarrow S \subseteq T \Rightarrow \overline{C}(X \backslash S) \cup \overline{C}(X \backslash T) = X \\
&\quad \text{or } \overline{C}(X \backslash S) \supseteq \overline{C}(X \backslash T) \\
&\Leftrightarrow X \backslash S \supseteq X \backslash T (\text{or } S \subseteq T) \Rightarrow \overline{C}(S) \cup \overline{C}(T) = X \\
&\quad \text{or } \overline{C}(T) \supseteq \overline{C}(S) \\
&\Leftrightarrow \overline{C} \text{ satisfies condition } \beta.
\end{aligned}$$

(2) C satisfies condition $\alpha \Leftrightarrow \overline{C}$ satisfies condition α. We have for $\forall S, T \in \mathscr{P}^*(X)$,

$$\begin{aligned}
C \text{ satisfies condition } \alpha &\Leftrightarrow S \subseteq T \Rightarrow C(T) \cap S \subseteq C(S) \\
&\Leftrightarrow S \subseteq T \Rightarrow (X \backslash \overline{C}(X \backslash T)) \cap S \subseteq X \backslash \overline{C}(X \backslash S) \\
&\Leftrightarrow S \subseteq T \Rightarrow X \backslash (\overline{C}(X \backslash T) \cup (X \backslash S)) \subseteq X \backslash \overline{C}(X \backslash S) \\
&\Leftrightarrow S \subseteq T \Rightarrow \overline{C}(X \backslash T) \cup (X \backslash S) \supseteq \overline{C}(X \backslash S) \\
&\Leftrightarrow X \backslash S \supseteq X \backslash T (\text{or } S \subseteq T) \Rightarrow \overline{C}(S) \cup T \supseteq \overline{C}(T) \\
&\Leftrightarrow \overline{C} \text{ satisfies condition } \alpha.
\end{aligned}$$

(3) C satisfies condition $\gamma \Leftrightarrow \overline{C}$ satisfies condition γ. We have for $\forall S, T \in \mathscr{P}^*(X)$,

$$\overline{C} \text{ satisfies condition } \gamma \Leftrightarrow C(S) \cap C(T) \subseteq C(S \cup T)$$
$$\Leftrightarrow (X \backslash \overline{C}(X \backslash S)) \cap (X \backslash \overline{C}(X \backslash T)) \subseteq X \backslash \overline{C}(X \backslash S \cup T)$$
$$\Leftrightarrow X \backslash (\overline{C}(X \backslash S) \cup \overline{C}(X \backslash T)) \subseteq X \backslash \overline{C}(X \backslash S \cup T)$$
$$\Leftrightarrow \overline{C}(X \backslash S) \cup \overline{C}(X \backslash T) \supseteq \overline{C}(X \backslash S \cup T)$$
$$\Leftrightarrow \overline{C}(X \backslash S) \cup \overline{C}(X \backslash T) \supseteq \overline{C}((X \backslash S) \cap X \backslash T))$$
$$\Leftrightarrow \overline{C}(S) \cup \overline{C}(T) \supseteq \overline{C}(S \cap T)$$
$$\Leftrightarrow \overline{C} \text{ satisfies condition } \gamma.$$

(4) C satisfies Arrow $\Leftrightarrow \overline{C}$ satisfies Arrow. We have for $\forall S, T \in \mathscr{P}^*(X)$,

$$C \text{ satisfies Arrow} \Leftrightarrow S \subseteq T \Rightarrow S \cap C(T) = \emptyset \text{ or } S \cap C(T) = C(S)$$
$$\Leftrightarrow S \subseteq T \Rightarrow S \cap (X \backslash \overline{C}(X \backslash T)) = \emptyset$$
$$\text{or } S \cap (X \backslash \overline{C}(X \backslash T)) = X \backslash \overline{C}(X \backslash S)$$
$$\Leftrightarrow S \subseteq T \Rightarrow X \backslash ((X \backslash S) \cup \overline{C}(X \backslash T)) = \emptyset$$
$$\text{or } X \backslash ((X \backslash S) \cup \overline{C}(X \backslash T)) = X \backslash \overline{C}(X \backslash S)$$
$$\Leftrightarrow S \subseteq T \Rightarrow (X \backslash S) \cup \overline{C}(X \backslash T) = X$$
$$\text{or } (X \backslash S) \cup \overline{C}(X \backslash T) = \overline{C}(X \backslash S)$$
$$\Leftrightarrow X \backslash S \supseteq X \backslash T (\text{or } S \subseteq T) \Rightarrow T \cup \overline{C}(S) = X$$
$$\text{or } T \cup \overline{C}(S) = \overline{C}(T)$$
$$\Leftrightarrow \overline{C} \text{ satisfies Arrow.}$$

(5) C satisfies WARP $\Leftrightarrow \overline{C}$ satisfies WARP. We have for $\forall S, T \in \mathscr{P}^*(X)$,

$$C \text{satisfies WARP} \Leftrightarrow x \in C(S), \ y \in S \backslash C(S), \text{ and } y \in C(T) \Rightarrow x \notin T$$
$$\Leftrightarrow x \in X \backslash \overline{C}(X \backslash S), y \in S \backslash (X \backslash \overline{C}(X \backslash S),$$
$$\text{and } y \in X \backslash C(X \backslash T) \Rightarrow x \notin T$$
$$\Leftrightarrow x \notin \overline{C}(X \backslash S), y \in S, y \in \overline{C}(X \backslash S),$$
$$\text{and } y \notin C(X \backslash T) \Rightarrow x \notin T$$
$$\Leftrightarrow x \notin \overline{C}(S), \ y \in \overline{C}(S) \backslash S, \text{ and } y \notin \overline{C}(T) \Rightarrow x \in T$$
$$(\text{interchanging} S \text{ and } X \backslash S \text{ as well as } T \text{ and} X \backslash T)$$
$$\Leftrightarrow \overline{C} \text{ satisfies WARP.}$$

(6) C satisfies $PI \Leftrightarrow \overline{C}$ satisfies Pareto Independence. We have for $\forall S, T \in \mathscr{P}^*(X)$,

$$C \text{ satisfies } PI \Leftrightarrow C(S \cup T) = C(C(S) \cup C(T))$$
$$\Leftrightarrow X \backslash \overline{C}(X \backslash S \cup T) = X \backslash \overline{C}(X \backslash (C(S) \cup C(T)))$$
$$\Leftrightarrow \overline{C}(X \backslash S \cup T) = \overline{C}(X \backslash (C(S) \cup C(T)))$$

$$\Leftrightarrow \overline{C}((X\backslash S) \cap (X\backslash T)) = \overline{C}((X\backslash C(S)) \cap (X\backslash C(T)))$$

$$\Leftrightarrow \overline{C}(S \cap T) = \overline{C}((X - C(X\backslash S)) \cap (X\backslash C(X\backslash T)))$$

$$\Leftrightarrow \overline{C}(S \cap T) = \overline{C}(\overline{C}(S) \cap \overline{C}(T))$$

$$\Leftrightarrow \overline{C} \text{ satisfies Pareto Independence.}$$

Let $\mathscr{L}$ denote the set of all lower choice functions on X and $\mathscr{U}$ the set of all upper choice functions on X. Define $\cap$ and $\cup$ on $\mathscr{L}$ as follows: $\forall C_1, C_2 \in \mathscr{L}, (C_1 \cap C_2)(S) = C_1(S) \cap C_2(S)$ and $(C_1 \cup C_2)(S) = C_1(S) \cup C_2(S) \forall S \in \mathscr{P}^*(X)$. Define $C_0 : \mathscr{P}^*(X) \to \mathscr{P}^*(X)$ by $\forall S \in \mathscr{P}^*(X), C_0(S) = S$. Then $(\mathscr{L}, \cap)$ is a commutative monoid with identity C_0, $(\mathscr{L}, \cup)$ is a commutative semigroup, and both distributive laws hold with respect to $\cap$ and $\cup$. Define $\cup$ and $\cap$ on $\mathscr{U}$ in a similar manner. Define $\overline{C_0} : \mathscr{P}^-(X) \to \mathscr{P}^-(X)$ by $\forall S \in \mathscr{P}^-(X), \overline{C_0}(S) = S$. Then $(\mathscr{U}, \cup)$ is a commutative monoid with identity $\overline{C_0}$, $(\mathscr{U}, \cap)$ is a commutative semigroup, and both distributive laws hold with respect to $\cup$ and $\cap$. Define $\subseteq$ on $\mathscr{L}$ by $\forall C_1, C_2 \in \mathscr{L}, C_1 \subseteq C_2$ if and only if $\forall S \in \mathscr{P}^*(X), C_1(S) \subseteq C_2(S)$. Define $\subset$ on $\mathscr{L}$ by $\forall C_1, C_2 \in \mathscr{L}, C_1 \subset C_2$ if and only if $C_1 \subseteq C_2$ and $\exists S \in \mathscr{P}^*(X), C_1(S) \subset C_2(S)$. Define $\subseteq$ and $\subset$ on $\mathscr{U}$ in a similar manner.

In the following result, we show that the algebraic structures $(\mathscr{L}, \cap, \cup)$ and $(\mathscr{U}, \cup, \cap)$ are isomorphic.

Proposition 7.25 *Define the relation f of $\mathscr{L}$ into $\mathscr{U}$ by $\forall C \in \mathscr{L}, f(C) = \overline{C}$, where $\forall S \in \mathscr{P}^-(X), \overline{C}(S) = X\backslash C(X\backslash S))$. Then f is a one-to-one function of $\mathscr{L}$ onto $\mathscr{U}$ such that $\forall C_1, C_2 \in \mathscr{L}, f(C_1 \cup C_2) = f(C_1) \cap f(C_2), f(C_1 \cap C_2) = f(C_1) \cup f(C_2)$, and $C_1 \subset C_2$ if and only if $f(C_1) \supset f(C_2)$.*

Proof Let $S \in \mathscr{P}^+(X)$. Since $X\backslash S \in \mathscr{P}^-(X), \overline{C}$ is defined on $\mathscr{P}^-(X)$. Since $S \neq \emptyset, X\backslash S \neq X$. Since also $C(S) \subseteq S, \overline{C}(X\backslash S) \supseteq X\backslash S$. From this it follows that $\overline{C} \in \mathscr{U}$. Let $C_1, C_2 \in \mathscr{L}$. Then

$$\begin{aligned}
C_1 = C_2 &\Leftrightarrow \forall S \in \mathscr{P}^+(X), C_1(S) = C_2(S) \\
&\Leftrightarrow \forall S \in \mathscr{P}^+(X), C_1(X\backslash S) = C_2(X\backslash S) \\
&\Leftrightarrow \forall S \in \mathscr{P}^+(X), X\backslash C_1(X\backslash S) = X\backslash C_2(X\backslash S) \\
&\Leftrightarrow \forall S \in \mathscr{P}^-(X), \overline{C_1}(S) = \overline{C_2}(S) \\
&\Leftrightarrow \overline{C_1} = \overline{C_2} \\
&\Leftrightarrow f(C_1) = f(C_2).
\end{aligned}$$

Therefore, f is singled-valued and one-to-one.

Let $\overline{C} \in \mathscr{U}$. Define $C : \mathscr{P}^+(X) \to \mathscr{P}^+(X)$ by $\forall S \in \mathscr{P}^+(X), C(S) = X\backslash\overline{C}(X\backslash S)$. Then $C \in \mathscr{L}$ since $C(S) \subseteq S$ and C is defined on $\mathscr{P}^+(X)$. Now $\forall S \in \mathscr{P}^-(X)$

$$\begin{aligned}
f(C)(S) &= X\backslash C(X\backslash S) \\
&= X\backslash(X\backslash\overline{C}(X\backslash(X\backslash S))
\end{aligned}$$

$$= X \backslash (X \backslash \overline{C}(S))$$
$$= \overline{C}(S) \ .$$

Hence $f(C) = \overline{C}$ and so f maps $\mathscr{L}$ onto $\mathscr{U}$.

We have $\forall S \in \mathscr{P}^-(X)$

$$
\begin{aligned}
(f(C_1) \cap f(C_2))(S) &= f(C_1)(S) \cap f(C_2)(S) \\
&= (X \backslash C_1(X \backslash S)) \cap (X \backslash C_2(X \backslash S)) \\
&= X \backslash (C_1(X \backslash S) \cup C_2(X \backslash S)) \\
&= X \backslash (C_1 \cup C_2)(X \backslash S) \\
&= f(C_1 \cup C_2)(S) \ .
\end{aligned}
$$

Thus $f(C_1) \cap f(C_2) = f(C_1 \cup C_2)$. Similarly, $f(C_1) \cup f(C_2) = f(C_1 \cap C_2)$. We also have

$$
\begin{aligned}
C_1 \subset C_2 &\Leftrightarrow C_1 \subseteq C_2 \text{ and } \exists S \in \mathscr{P}^+(X), C_1(S) \subset C_2(S) \\
&\Leftrightarrow C_1 \subseteq C_2 \text{ and } \exists S \in \mathscr{P}^+(X), X \backslash C_1(S) \supset X \backslash C_2(S) \\
&\Leftrightarrow C_1 \subseteq C_2 \text{ and } \exists S \in \mathscr{P}^-(X), X \backslash C_1(X \backslash S) \supset X \backslash C_2(X \backslash S) \\
&\Leftrightarrow \overline{C_1} \supseteq \overline{C_2} \text{ and } \exists S \in \mathscr{P}^-(X), \overline{C_1}(S) \supset \overline{C_2}(S) \\
&\Leftrightarrow \overline{C_1} \supset \overline{C_2} \Leftrightarrow f(C_1) \supset f(C_2).
\end{aligned}
$$

Hence the desired result holds.

Corollary 7.26 *Define the relation g of $\mathscr{U}$ into $\mathscr{L}$ by $\forall \overline{C} \in \mathscr{U}$, $g(\overline{C}) = C$, where $\forall S \in \mathscr{P}^*(X)$, $C(S) = X \backslash \overline{C}(X \backslash S))$. Then g is a one-to-one function of $\mathscr{U}$ onto $\mathscr{L}$ such that $\forall \overline{C_1}, \overline{C_2} \in \mathscr{U}$, $g(\overline{C_1} \cap \overline{C_2}) = g(\overline{C_1}) \cup g(\overline{C_2})$, $g(\overline{C_1} \cup \overline{C_2}) = g(\overline{C_1}) \cap g(\overline{C_2})$, and $\overline{C_1} \subset \overline{C_2}$ if and only if $g(\overline{C_1}) \supset g(\overline{C_2})$. Furthermore, $f(g(\overline{C})) = \overline{C}$ for $\forall \overline{C} \in \mathscr{U}$ and $g(f(C)) = C$ for $\forall C \in \mathscr{L}$.*

Proof $g = f^{-1}$.

7.3.4 Comments and Future Research

Let C be a choice function on X and $\overline{C}$ the corresponding upper choice function on X. Let R be a relation on X that rationalizes C. Let $S \in \mathscr{P}^+(X) \cap \mathscr{P}^-(X)$. Then

$$
\begin{aligned}
\overline{C}(S) &= X \backslash C(X \backslash S) \\
&= S \cup (C(X \backslash S) \backslash S) \\
&= C(S) \cup (S \backslash C(S)) \cup (C(X \backslash S) \backslash S) \\
&= M(R, S) \cup (S \backslash M(R, S)) \cup (C(X \backslash S) \backslash S)
\end{aligned}
$$

and $\{C(S), (S\backslash C(S)), C(X\backslash S)\backslash S\}$ and $\{M(R, S), S\backslash M(R, S), C(X\backslash S)\backslash S)\}$ are partitions of X.

We now show that C and $\overline{C}$ cannot be defined nontrivially in terms of an equivalence relation E on X as is the case in rough set theory. Suppose C is a choice function on X such that $\forall S \in \mathscr{P}^+(X)$, $C(S) = \{x \in X \mid [x] \subseteq S\}$ and $\overline{C}(S) = \{x \in X \mid [x] \cap S \neq \emptyset\}$, where $[x]$ is the equivalence class of E induced by x. Then for $s \in X$, $C(\{s\}) = \{x \in X \mid [x] \subseteq \{s\}\} = \{s\}$ and so $\forall s \in X$, $[s] = \{s\}$. Thus if rough set theory is to be applied, the use of an equivalence relation should be replaced with that of a cover. Results concerning the development of rough set theory based on covers can be found in Orlowska (1986), Pomylaka (1987), Slowinski and Vanderpooten (1995), Wasilewski (1990), Wasilewski and Vigneron (1997), Wybraniec-Skardowska (1989). Applications of these approaches to topology, (fuzzy) abstract algebra, (fuzzy) directed graphs, (fuzzy) finite state machines, modal logic, and interval structures can be found in Kuroki and Mordeson (1997a, b), Mordeson (1999a, b), Malik and Mordeson (2002), Yao (1996), Yao and Lin (1996), Yao (1998).

References

Austen-Smith, D., Banks, J.: Positive Political Theory I. The University of Michigan Press, Ann Arbor (1999)

Austen-Smith, D., Banks, J.: Positive Political Theory II. The University of Michigan Press, Ann Arbor (2005)

Fedrizzi, M., Kacprzyk, J., Nurmi, H.: How different are social choice functions: a rough sets approach. Qual. Quant. **30**(1), 87–99 (1996). http://dx.doi.org/10.1007/BF00139836

Kuroki, N., Mordeson, J.N.: Structure of rough sets and rough groups. J. Fuzzy Math. **5**(1), 183–191 (1997a)

Kuroki, N., Mordeson, J.N.: Successor and source functions. J. Fuzzy Math. **5**, 173–182 (1997b)

Malik, D.S., Mordeson, J.N.: Structure of upper and lower approximation spaces of infinite sets. In: Lin, T.Y., Yao, Y.Y., Zadeh, L.A. (eds.) Data Mining, Rough Sets and Granular Computing, pp. 461–473. Physica-Verlag GmbH, Heidelberg, Germany (2002). http://dl.acm.org/citation.cfm?id=783032.783055

Mordeson, J.N.: Algebraic properties of lower approximation spaces. J. Fuzzy Math. **7**, 631–637 (1999a)

Mordeson, J.N.: Algebraic properties of spaces in rough fuzzy set theory. In: Fuzzy Information Processing Society, 1999. NAFIPS 18th International Conference of the North American, pp. 56–59 (1999b)

Mordeson, J.N., Bhutani, K., Clark, T.D.: The rationality of fuzzy choice functions. New Math. Nat. Comput. **4**, 309–327 (2008)

Orlowska, E.: Semantic analysis of inductive reasoning. Theor. Comput. Sci. **43**, 81–89 (1986)

Pawlak, Z.: Rough Sets: Theoretical Aspects of Reasoning About Data. Kluwer, Boston (1991)

Pomylaka, J.A.: Approximation operations in approximation space. Bull. Pol. Acad. Sci. Math. **35**, 653–662 (1987)

Slowinski, R., Vanderpooten, D.: Similarity relation as a basis for rough approximations. In: Wang, P.P. (ed.) Advances in Machine Intelligence and Soft Computing, pp. 17–33 (1995)

Wasilewski, A.: Conditional knowledge representation systems—model for an implementation. Bull. Pol. Acad. Sci. Math. **37**(1–6), 63–69 (1990)

Wasilewski, A., Vigneron, L.: On generalized rough sets. In: Proceedings of the 5th Workshop on Rough Sets and Soft Computing RSSC'97. 3rd Joint Conference on Information Science (1997)

Wybraniec-Skardowska, U.: On a generalization of approximation space. Bull. Pol. Acad. Sci. Math. **37**, 51–61 (1989)

Yao, Y.Y., Lin, T.Y.: Generalization of rough sets using modal logics. Intell. Autom. Soft Comput. Int. J. **2**, 103–120 (1996)

Yao, Y.: Two views of the theory of rough sets in finite universes. Int. J. Approximate Reasoning **15**(4), 291–317 (1996). http://www.sciencedirect.com/science/article/pii/S0888613X96000710, (rough Sets)

Yao, Y.: Relational interpretations of neighborhood operators and rough set approximation operators. Inf. Sci. **111**(1–4), 239–259 (1998). http://www.sciencedirect.com/science/article/pii/S0020025598100063

Appendix A
Full Results

This book compares six government formation models: the conventional Median Voter model, the conventional proximity model, the fuzzy single-dimension Pareto set model, the fuzzy single-dimension maximal set model, the fuzzy weighted single-dimension maximal set model, and the fuzzy two-dimension maximal set model. Each of the models presented in this paper minimally offer an alternative to traditional Euclidean models when attempting to predict coalition formation. In most cases, the one-dimensional and two-dimensional fuzzy models report opportunities for equal or improved prediction rates over the traditional euclidean median voter model. In addition, the fuzzy weighted government formation model modestly improves the prediction rates of minimal-winning coalitions and super-majority coalitions. However, no single model offers the best prediction rate across all models for each coalition type, which still leaves something to be desired. That being said, the fuzzy weighted one-dimensional model does offer the best total prediction rate at 60.2%. Below, are the results of each model addressed in this paper. Models which offer the best prediction rate for a coalition type are *emphasized*.

The conventional proximity model offers the best total prediction rate for single party minority coalitions at 79.55%. The fuzzy two-dimensional model best predicts minority coalitions with a rate of 62.5%. All models except the fuzzy weighted one-dimensional model correctly predict single party majority at 100%. Minimal winning coalitions and supermajorities are both best predicted by the fuzzy weighted one-dimensional Fuzzy model at rates of 68 and 54%, respectively. In sum the model with the best prediction rate differs when a researcher is interested in a specific coalition type. However, if government formation–regardless of coalition type–is the focus of the research agenda, then the model which best correctly predicts government formation is not one of the conventional models, but the fuzzy weighted one-dimensional model (Tables A.1 and A.2).

P. C. Casey et al., *Fuzzy Social Choice Models*, Studies in Fuzziness
and Soft Computing 318, DOI: 10.1007/978-3-319-08248-6,
© Springer International Publishing Switzerland 2014

Table A.1 Government formation results, part one

Government (# of cases)	First-level	Other-level	No prediction	Percent correct	# of levels
Median voter					
Single party minority (44)	23 (52.27 %)	21 (47.73 %)	N/A	23 (52.27 %)	N/A
Minority coalition (27)	0 (0.00 %)	27 (100 %)	N/A	0 (0.00 %)	N/A
Single party majority (23)	23 (100 %)	0 (0.00 %)	N/A	23 (100 %)	N/A
Minimal winning coalition (89)	0 (0.00 %)	89 (100 %)	N/A	0 (0.00 %)	N/A
Supermajority (52)	0 (0.00 %)	52 (100 %)	N/A	0 (0.00 %)	N/A
All cases (235)	46 (19.57 %)	189 (80.43 %)	N/A	46 (19.57 %)	N/A
Crisp proximity					
Single party minority (44)	35 (79.55 %)	9 (20.45 %)	N/A	35 (79.55 %)	N/A
Minority coalition (27)	8 (29.63 %)	19 (70.37 %)	N/A	8 (29.63 %)	N/A
Single party majority (23)	23 (100 %)	0 (0.00 %)	N/A	23 (100 %)	N/A
Minimal winning coalition (89)	26 (29.21 %)	63 (70.79 %)	N/A	26 (29.21 %)	N/A
Supermajority (52)	6 (11.54 %)	46 (88.46 %)	N/A	6 (11.54 %)	N/A
All cases (235)	98 (41.7 %)	137 (58.30 %)	N/A	98 (41.7 %)	N/A
Casey 1-D fuzzy					
Single party minority (45)	24 (53.33 %)	0 (0.00 %)	3 (6.67 %)	24 (53.33 %)	2.40
Minority coalition (26)	9 (34.62 %)	5 (19.23 %)	1 (3.85 %)	14 (53.8 %)	2.60
Single party majority (21)	21 (100 %)	0 (0.00 %)	0 (0.00 %)	21 (100 %)	2.70
Minimal winning coalition (87)	25 (28.74 %)	18 (20.69 %)	13 (14.94 %)	43 (49.2 %)	2.30
Supermajority (52)	6 (11.54 %)	11 (21.15 %)	5 (9.62 %)	17 (32.69 %)	2.50
All cases (231)	88 (36.81 %)	34 (14.72 %)	22 (9.52 %)	119 (51.5 %)	2.40

Table A.2 Government formation results, part two

Gibilisco 1-D fuzzy

Government type (#of cases)	First level	Other level	No prediction	Percent correct	# of levels
Single party minority (17)	11 (64.71 %)	0 (0.0 %)	4 (23.53 %)	11 (64.71 %)	2.9
Minority coalition (27)	7 (25.93 %)	5 (18.52 %)	5 (18.52 %)	12 (44.44 %)	3.0
Single party majority (22)	22 (100 %)	0 (0.00 %)	0 (0.00 %)	22 (100 %)	3.2
Minimal winning coalition (89)	21 (23.60 %)	17 (19.10 %)	18 (20.22 %)	38 (42.70 %)	2.9
Supermajority (52)	6 (11.54 %)	11 (21.15 %)	9 (17.31 %)	17 (32.69 %)	3.0
All cases (207)	75 (36.62 %)	30 (14.49 %)	36 (17.39 %)	100 (48.31 %)	3.0

Gibilisco 2-D fuzzy

Government type (# of cases)	First level	Second level	Other level	No prediction	Percent correct	# of levels
Single party minority (18)	12 (66.67 %)	1 (5.56 %)	0 (0.00 %)	4 (22.22 %)	13 (72.22 %)	2.8
Minority coalition (8)	3 (37.50 %)	1 (12.50 %)	1 (12.50 %)	2 (25.00 %)	5 (62.50 %)	3.3
Single party majority (11)	11 (100.00 %)	0 (0.00 %)	0 (0.00 %)	0 (0.00 %)	11 (100.00 %)	3.2
Minimal winning coalition (44)	15 (34.09 %)	3 (6.82 %)	1 (2.27 %)	15 (34.09 %)	19 (43.18 %)	3.0
Supermajority (16)	2 (12.50 %)	3 (18.75 %)	2 (12.50 %)	3 (18.75 %)	7 (43.75 %)	2.6
All cases (97)	43 (44.33 %)	8 (8.25 %)	4 (4.12 %)	24 (24.74 %)	55 (56.70 %)	2.8

Fuzzy weighted 1-D

Government type (# of Cases)	First level	Second level	Third level	Fourth level	Other level	Percent correct
Single party minority (45)	5 (11.11 %)	11 (24.44 %)	5 (11.11 %)	4 (8.89 %)	20 (44.44 %)	25 (55.55 %)
Minority coalition (28)	6 (21.42 %)	2 (7.14 %)	1 (3.57 %)	2 (7.1 %)	17 (6.07 %)	11 (39.20 %)
Single party majority (26)	10 (38.46 %)	7 (26.92 %)	1 (3.85 %)	1 (3.85 %)	7 (26.92 %)	19 (73.10 %)
Minimal winning coalition (100)	23 (23.00 %)	32 (32.00 %)	7 (7.00 %)	6 (6.00 %)	32 (32.00 %)	68 (68.00 %)
Supermajority coalition (50)	13 (26.00 %)	4 (8.00 %)	8 (16.00 %)	2 (4.00 %)	23 (23.00 %)	27 (54.00 %)
All cases (249 total)	57 (22.89 %)	56 (22.49 %)	22 (8.84 %)	15 (6.02 %)	99 (39.76 %)	150 (60.20 %)

Appendix B
Selected Definitions and Proofs

Let, $\sigma_i(x) \in [0, 1]$.

Definition B.1

$$\rho_i(x, y) = \begin{cases} [\sigma_i(x) - \sigma_i(y) + r] \wedge 1 & \text{if } \sigma_i(x) \geq \sigma_i(y) \\ 1 - [(\sigma_i(y) - \sigma_i(x) + 1 - r) \wedge 1] & \text{if } \sigma_i(x) < \sigma_i(y), \end{cases}$$

where $r \in [0, 1]$.

Proposition B.2 $\sigma_i(x) > \sigma_i(y)$, then $\rho_i(x, y) > \rho_i(y, x)$.

Proof Assume $\sigma_i(x) > \sigma_i(y)$. Suppose $\rho_i(x, y) \leq \rho_i(y, x)$. Then by substitution

$$(\sigma_i(x) - \sigma_i(y) + r) \wedge 1 \leq 1 - [(\sigma_i(y) - \sigma_i(x) + 1 - r) \wedge 1]$$

or

$$(\sigma_i(y) + r) \wedge 1 > (1 - \sigma_i(y) + \sigma_i(x) - 1 + r) \vee 1 - 1].$$

Simplifying

$$(\sigma_i(x) - \sigma_i(y) + r) \wedge 1 \leq -\sigma_i(y) + \sigma_i(x) + r \vee 0.$$

However this is a contradiction because $\sigma_i(x) > \sigma(y)$ and

$$\sigma_i(x) - \sigma_i(y) > 0 > -\sigma_i(y) + \sigma_i(y) + \sigma_i(x).$$

Hence, $\rho_i(x, y) > \rho_i(y, x)$.

Definition B.3

$$M_{G_i}(\rho_i, \sigma_i)(x) = \sigma_i(x) \wedge$$
$$\bigwedge \left\{ \bigvee \{t \in [0, 1] \mid \sigma_i(y) \wedge \rho_i(y, x) \wedge t \leq \rho_i(x, y)\} \mid y \in \text{supp}(\sigma_i) \right\}$$

P. C. Casey et al., *Fuzzy Social Choice Models*, Studies in Fuzziness
and Soft Computing 318, DOI: 10.1007/978-3-319-08248-6,
© Springer International Publishing Switzerland 2014

Proposition B.4 $\sigma_i(x) = 1 \Leftrightarrow M_{G_i}(\rho_i, \sigma_i)(x) = 1$.

Proof Suppose $\sigma_i(x) = 1$. If $\sigma_i(x) = 1$, then

$$M_{G_i}(\rho_i, \sigma_i)(x) = 1 \wedge$$
$$\bigwedge \left\{ \bigvee \{t \in [0, 1] \mid 1 \wedge \rho_i(y, x) \wedge t \leq \rho(x, y)\}\} \mid y \in \text{Supp}(\sigma_i) \right\}.$$

For all $y \in \text{Supp}(\sigma_i)$, $\rho_i(x, y) \geq r \geq \rho_i(y, x)$, by Definition B.1. Thus, $\vee t = 1$ for all $y \in \text{Supp}(\sigma_i)$ because $\rho_i(y, x) \leq \rho_i(x, y)$. Hence, $M_{G_i}(\rho_i, \sigma_i)(x) = 1$.

Now suppose $M_{G_i}(\rho_i, \sigma_i)(x) = 1$. Since, by Definition B.3, $M_{G_i}(\rho_i, \sigma_i)(x) = \sigma_i(x) \wedge t$ for some $t \in [0, 1]$, then $\sigma_i(x) = 1$.

Proposition B.5 $M_{G_i}(\rho_i, \sigma_i)(x) \leq \sigma_i(x)$.

Proof The proof is intuitive. $M_{G_i}(\rho_i, \sigma_i)(x) = \sigma_i(x) \wedge t$. Thus, $M_{G_i}(\rho_i, \sigma_i)(x)$ can be no greater than $\sigma_i(x)$.

Proposition B.6 $r = 1 \Rightarrow M_{G_i}(\rho_i, \sigma_i)(x) = \sigma_i(x)$.

Proof Suppose $r = 1$. Solving for $M_{G_i}(\rho_i, \sigma_i)(x)$, there are two cases for $\rho(y, x)$ and $\rho(x, y)$, either $\sigma_i(y) > \sigma_i(x)$ or $\sigma_i(x) \geq \sigma_i(y)$.

Assume $\sigma_i(x) \geq \sigma_i(y)$. Then, by Definition B.1, $\rho_i(x, y) = 1$ for all $y \in supp(\sigma_i)$ because $(\sigma_i(x) - \sigma_i(y) + 1) \geq 1$. Since, by Definition B.1, $1 \geq \rho_i(x, y)$ and $\rho_i(y, x), \rho_i(x, y) \geq \rho_i(y, x)$ because $\rho_i(x, y) = 1$. Thus, by Definition B.3, $\rho_i(y, x) \leq \rho_i(x, y)$ and $\bigvee \{t \in [0, 1] = 1\}$ when $\sigma_i(x) \geq \sigma_i(y)$. Since $\bigvee \{t \in [0, 1] = 1\}$,

$$M_{G_i}(\rho_i, \sigma_i)(x) = \sigma_i(x) \wedge 1$$
$$= \sigma_i(x)$$

because $\sigma_i(x) \in [0, 1]$.

Assume $\sigma_i(y) > \sigma_i(x)$. If $\sigma_i(y) > \sigma_i(x)$, then $\rho_i(y, x) = 1$, for all $y \in supp(\sigma_i)$. Then $\rho(x, y) = 1 - [(\sigma_i(y) - \sigma_i(x) + -1) \wedge 1]$. Since $r = 1$, $\rho_i(x, y) \geq \sigma_i(x)$ because

$$1 - (\sigma_i(y) - \sigma_i(x)) = 1 + \sigma_i(x) - \sigma_i(y)$$

and $\sigma_i(y) \in [0, 1]$. Thus, by Definition B.3, $\sigma_i(x) \leq \rho_i(x, y)$ and $\bigvee \{t \in [0, 1] = 1$. Thus,

$$M_{G_i}(\rho_i, \sigma_i)(x) = \sigma_i(x) \wedge 1$$
$$= \sigma_i(x)$$

because $\sigma_i(x) \in [0, 1]$.

Definition B.7 A fuzzy preference relation, f, is *reflexive* if for all $x \in X$, $f(x, x) = 1$.

Proposition B.8 ρ_i *is reflexive if and only if* $r = 1$.

Proof The proof is intuitive.

$$[(\sigma_i(x) - \sigma_i(x) + r \wedge 1] = [r \wedge 1]$$
$$= c.$$

Definition B.9 A fuzzy preference relation, f, is *complete* if $f(x, y) > 0$ or $f(y, x) > 0$ for all $x, y \in X$.

Proposition B.10 *If* $r = 1$, *then* ρ_i *is complete.*

Proof Assume $r = 1$. For any $x, y \in X$, either $\sigma_i(x) \geq \sigma_i(y)$ or $\sigma_i(y) \geq \sigma_i(x)$. Assume the former. Then

$$\rho_i(x, y) = [(\sigma_i(x) - \sigma_i(y) + 1) \wedge 1]$$
$$= 1,$$

by Definition B.1.
   Now assume the latter. Then

$$\rho_i(y, x) = [(\sigma_i(x) - \sigma_i(y) + 1) \wedge 1]$$
$$= 1,$$

by Definition B.1. In either case, $\rho_i(x, y)$ or $\rho_i(y, x) > 0$ for any $x, y \in X$.

Index

Printed in the United States
By Bookmasters